PEARLS OF WISDOM SERIES

EMT-Basic

PEARLS OF WISDOM

SECOND EDITION

Guy H. Haskell, PhD, NREMT-P
Director, Emergency Medical and Safety Services Consultants
Bloomington, Indiana
Fire Fighter, Benton Township Volunteer Fire Department
Unionville, Indiana

Robert C. Krause, EMT-P
Captain, Toledo Fire and Rescue
Toledo, Ohio

JONES AND BARTLETT PUBLISHERS
Sudbury, Massachusetts
BOSTON TORONTO LONDON SINGAPORE

Jones and Bartlett Publishers
World Headquarters
Jones and Bartlett Publishers
40 Tall Pine Drive
Sudbury, MA 01776
978-443-5000
info@jbpub.com
www.EMSzone.com

Jones and Bartlett Publishers Canada
6339 Ormindale Way
Mississauga, ON L5V 1J2
Canada

Jones and Bartlett Publishers International
Barb House, Barb Mews
London W6 7PA
United Kingdom

Production Credits
Chief Executive Officer: Clayton E. Jones
Chief Operating Officer: Donald W. Jones, Jr.
President, Higher Education and Professional Publishing: Robert W. Holland, Jr.
V.P., Sales and Marketing: William J. Kane
V.P., Production and Design: Anne Spencer
V.P., Manufacturing and Inventory Control: Therese Connell
Publisher, Public Safety: Kimberly Brophy
Director of Marketing: Alisha Weisman
Aquisitions Editor, EMS: Christine Emerton
Editorial Assistant: Adrienne Zicht
Production Editor: Susan Schultz
Composition: N.K. Graphics
Text and Cover Design: Anne Spencer
Printing and Binding: Courier Corporation
Cover Printing: Lehigh Press

Jones and Bartlett's books and products are available through most bookstores and online booksellers. To contact Jones and Bartlett Publishers directly, call 800-832-0034, fax 978-443-8000, or visit our website, www.jbpub.com.

Substantial discounts on bulk quantities of Jones and Bartlett's publications are available to corporations, professional associations, and other qualified organizations. For details and specific discount information, contact the special sales department at Jones and Bartlett via the above contact information or send an e-mail to specialsales@jbpub.com.

ISBN: 0-7637-4227-9

The procedures in this book are based on the most current recommendations of responsible medical sources. The publisher and author, however, make no guarantee as to, and assume no responsibility for, the correctness, sufficiency, or completeness of such information or recommendations. Other or additional safety measures may be required under particular circumstances.

This textbook is intended solely as a guide to the appropriate procedures to be employed when rendering emergency care to the sick and injured. It is not intended as a statement of the standards of care required in any particular situation, because circumstances and the patient's physical condition can vary widely from one emergency to another. Nor is it intended that this textbook shall in any way advise emergency personnel concerning legal authority to perform the activities or procedures discussed. Such local determinations should be made only with the aid of legal counsel.

Library of Congress Cataloging-in-Publication Data

Haskell, Guy H.
EMT-basic pearls of wisdom / Guy H. Haskell. -- 2nd ed.
p. ; cm.
Rev. ed. of: Emergency medical technician-basic. c1999.
ISBN 0-7637-4227-9 (pbk.)
1. Emergency medical services--Examinations, questions, etc.
2. Emergency medical technicians--Examinations, questions, etc.
I. Haskell, Guy H. Emergency medical technician-basic. II. Title.
[DNLM: 1. Emergency Medical Services--Examination Questions.
2. Emergency Medical Technicians--Examination Questions.
3. Emergencies--Examination Questions. WX 18.2 H349e 2007]
RA645.5.H37 2007
362.18076--dc22

2006021692

Printed in the United States of America
10 09 08 07 06 10 9 8 7 6 5 4 3 2 1

Contents

Introduction

Nothing in the world can take the place of persistence. Talent will not; nothing is more common than unsuccessful men with talent. Genius will not; unrewarded genius is almost a proverb. Education will not; the world is full of educated derelicts. Persistence and determination alone are omnipotent.
—Calvin Coolidge

Congratulations! *EMT-Basic Pearls of Wisdom* will help you learn some basic prehospital emergency medicine, prepare for state and national certification and refresher exams, or review material from the *U.S. Department of Transportation Emergency Medical Technician-Basic: National Standard Curriculum* (1994). It can also serve as a helpful adjunct for instructors conducting initial or refresher EMT-Basic training; it can serve as a short answer test bank, study guide, or review tool.

Pearls' unique format differs from all other review and test preparation texts. Let us begin, then, with a few words on purpose, format, limitations, and use.

The primary intent of *Pearls* is to serve as a study aid to improve performance on EMT-Basic certification and refresher examinations. With this goal in mind, the text is written in rapid-fire question and answer format. The student receives immediate gratification with a correct answer. Misleading or confusing multiple-choice "foils" are not provided, thereby eliminating the risk of assimilating erroneous information that made an impression. Another advantage of this format is that the student either knows or does not know the answer to a given question. This results in active learning, rather than the passive review of studying multiple-choice questions. Active learning goes beyond the desire to pass an exam; when administering epinephrine to a child in cardiac arrest, you must know instantly that you need to push .01 mg/kg, and there will be nobody to ask you if you want to push a) 1 mg, b) 0 .1 mg/kg, c) .01 mg/kg or d) none of the above.

The questions themselves often contain a pearl reinforced in association with the question and answer. Additional hooks are often attached to the answer in various forms, including mnemonics, evoked visual imagery, repetition, and humor. Additional information not requested in the question may be included in the answer. The same information is often sought in several different questions. Emphasis has been placed on evoking both trivia and key facts that are easily overlooked, are quickly forgotten, and yet somehow always seem to appear on certification exams.

Most questions have answers without explanations. This is done to enhance ease of reading and rate of learning. Explanations often occur in a later question or answer. It may happen that upon reading an answer the reader may think: "Hmm, why is that?" or, "Are you sure?" If this happens to you, go check! Truly assimilating these disparate facts into a framework of knowledge absolutely requires further reading in the surrounding concepts. Information learned as a response to seeking an answer to a particular question is much better retained than information that is passively read. Take advantage of this. Use *Pearls* with your preferred source texts handy and open, or, if you are reviewing on a train, a plane, or on camelback, mark questions for further investigation.

Pearls does have limitations. There are many conflicts among core sources in the field of emergency medical services. Not only are there discrepancies between texts, some texts have internal discrepancies, further confounding clarification of information. For better or worse, EMT-Basic examinations are based on the *U.S. Department of Transportation (DOT) Emergency Medical Technician-Basic: National Standard Curriculum* (1994). For the purposes of this review text, then, this curriculum serves as gospel.

The DOT curriculum represents "cook book" medicine. Formulated by a huge committee, it is also medicine by consensus. By its very nature, soon after publication many of the concepts will not represent the cutting edge of prehospital care. In addition, the curriculum contains lots of silly stuff, such as "depress the push-to-talk switch on the microphone before speaking." We know it's silly, you know it's silly, so don't spend a lot of time griping about it. State and national tests are based on this curriculum, so heave a deep sigh and get on with it.

With these limitations in mind, *Pearls* risks accuracy by aggressively pruning complex concepts down to the simplest kernel. The dynamic knowledge base and clinical practice of emergency medicine are not like that! The information taken as "correct" is that indicated in the DOT curriculum. Further, new research and practice occasionally deviates from that which likely represents the "right" answer for test purposes. In such cases we have selected the information that we believe is most likely "correct" for test purposes, that which most closely conforms to the DOT curriculum. This text is designed to maximize your score on a test. Refer to your most current sources of information, your mentors, and your medical director for direction on current practice.

Pearls is designed to be used, not just read. It is an interactive text. Use a sheet of paper to cover the answers; attempt all questions. A study method we strongly recommend is oral, group study, preferably over an extended meal (or pitchers). The mechanics of this method are simple and no one ever appears stupid. One person holds *Pearls*, with answers covered, and reads the question. Each person, including the reader, says "Check!" when he or she has an answer in mind. After everyone has "checked" in, someone states his or her answer. If this answer is correct, on to the next one. If not, another person states his or her answer, or the answer can be read. Usually, the person who "checks" in first gets the first shot at stating the answer. If this person is being a smarty-pants answer-hog, then others can take turns. Try it—it's almost fun!

Pearls is also designed to be re-used several times to allow, dare we use the word, memorization. If you are a pessimist, we suggest putting a check mark next to a question every time it is missed. If you have two check marks next to a question, forget this question! You will get it wrong on the exam! On the other hand, if you are an optimist, place a check mark when a question is answered correctly once; skip all questions with check marks thereafter. Utilize the scheme you prefer.

We welcome your comments, suggestions, and criticism. Great effort has been made to verify these questions and answers. There will be answers we have provided that are at variance with the answer you would prefer. Most often this is attributable to the reliance on the DOT curriculum. Please make us aware of any errata you find. We hope to make continuous improvements in future editions and would greatly appreciate any input with regard to format, organization, content, presentation, or about specific questions.

Study hard and good luck!

G.H.H. and R.C.K

Acknowledgments

Lead Editor

Guy H. Haskell, PhD, NREMT-P
Director, Emergency Medical and Safety Services Consultants
Bloomington, Indiana
Fire Fighter, Benton Township Volunteer Fire Department
Unionville, Indiana

Authors

Guy H. Haskell, PhD, NREMT-P
Director, Emergency Medical and Safety Services Consultants
Bloomington, Indiana
Fire Fighter, Benton Township Volunteer Fire Department
Unionville, Indiana

Robert C. Krause, EMT-P
Captain, Toledo Fire and Rescue
Toledo, Ohio

Contributing Author

Major Ray W. Burton, retired
Plymouth Police Academy
Plymouth, Massachusetts

Jones and Bartlett Publishers and the author would also like to acknowledge the contributors and reviewers of previous editions of **EMT-Basic Pearls of Wisdom**.

Section

1 Preparatory

Voices of Experience

My First Code

It was the first run of my first day of field clinicals for my paramedic class. I had been an EMT for about a year and a half, doing mostly interfacility transports. We had a dual response system with a medic truck with two medics, Ed and Tim (and me), and a BLS transport ambulance.

The very first call of the day was a cardiac arrest. I am reminded of it even 20 years later every time I pass the address. When we entered the apartment, our patient, a 50-something-year-old man, was lying on the couch. He was pulseless, apneic, and purple from the nipple-line up.

Jana, one of the EMTs, walked in, took one look at him, and in one swift move yanked him off the couch onto the floor and started CPR. I was in awe!

This was like mega-code practice, with the patient throwing every conceivable cardiac rhythm in the book at us. The medics couldn't get him intubated because his jaw was clenched, so we ventilated him with a bag-mask device throughout.

We finally got a pulse back, and we decided to move him onto the backboard to load him onto the cot. Being a student and still a novice, I remembered that our instructor had taught us that we should always talk to the patient, even if he is unconscious, because sometimes he can hear us and may even remember what we said afterward. So, being a conscientious little paramedic student, I said to the patient, "OK, sir, we're gonna move you onto the backboard now." He replied, "All right." Everyone froze. Mouths dropped open. We looked at each other over the patient. We were all thinking: Did he say that? I continued to talk to him. "Sir, how ya doin'?" "OK, I guess," he said.

By the time we had him in the squad, he was talking to us. The whole time I was thinking that this is the greatest job in the world! The first call of the first day, a save, and he is talking! I am an American hero! Of course, it was another 10 years or so until my next code spoke to me in the ambulance, and months would go by without anything more than band-aid calls.

That evening, high on my new profession, I stopped by the patient's room in the ICU to see how he was doing. He was sitting up watching TV. I said, "Hi, my name is Guy. I was one of the EMTs who brought you in today." Without looking away from the TV, he said, "Hi." No "Thanks for saving my life"; no happy hugs at the fire station—just a disinterested "Hi."

I learned several lessons that day. First, that some patients actually *can* hear you when they are unresponsive. Second, just because the more experienced medics don't do it doesn't mean it shouldn't be done. Third, sometimes you really can save a cardiac arrest—not often, but sometimes. Fourth, it is important to find satisfaction in your own work, even if you don't get thanked for it.

Guy Haskell, PhD, NREMT-P
Director, Emergency Medical and Safety Services Consultants
Bloomington, Indiana
Fire Fighter/Paramedic, Benton Township Volunteer Fire Department
Unionville, Indiana

Preparatory

"They do certainly give very strange, and newfangled, names to diseases."
—Plato

Section 1 covers the following subjects:

Lesson 1–1 Introduction to Emergency Medical Care

Familiarizes the EMT-Basic (EMT-B) candidate with the introductory aspects of emergency medical care. Topics covered include the emergency medical services system, roles and responsibilities of the EMT-B, quality improvement, and medical direction.

Lesson 1–2 Well-Being of the EMT-B

Covers the emotional aspects of emergency care, stress management, introduction to critical incident stress debriefing (CISD), scene safety, body substance isolation (BSI), personal protection equipment (PPE), and safety precautions that can be taken prior to performing the role of an EMT-B.

Lesson 1–3 Medical, Legal, and Ethical Issues

Explores the scope of practice, ethical responsibilities, advance directives, consent, refusals, abandonment, negligence, duty to act, confidentiality, and special situations such as organ donors and crime scenes. Medical, legal, and ethical issues are vital elements of the EMT-B's daily life.

Lesson 1–4 The Human Body

Enhances the EMT-B's knowledge of the human body. A brief overview of body systems, anatomy, physiology, and topographic anatomy will be given in this session.

Lesson 1–5 Baseline Vital Signs and SAMPLE History

Teaches how to assess and record a patient's vital signs and a SAMPLE history.

Lesson 1–6 Lifting and Moving Patients

Provides students with knowledge of body mechanics, lifting and carrying techniques, principles of moving patients, and an overview of equipment.

Does state, federal, or local government primarily regulate EMS?
State.

Which federal agency is responsible for recommending national standards?
The National Highway Traffic Safety Administration.

What are the two main benefits of a universal access number to EMS?

Provides trained staff to answer calls and minimizes delays.

What are the four levels of emergency medical technician training?

First Responder; EMT-Basic (EMT-B); EMT-Intermediate (EMT-I); EMT-Paramedic (EMT-P).

What is your primary responsibility as an EMT-B?

Personal safety and the safety of others.

What is the function of the initial assessment?

To identify and treat life-threatening problems.

Why do you perform a focused history and physical exam?

To identify and treat non-life-threatening problems.

What is the most important way you can prevent injury when lifting a patient?

Use proper body mechanics.

What is your first responsibility when arriving on a scene?

Scene safety.

You arrive on the scene of a motor vehicle crash. Utility wires are down around the vehicle. What do you do first?

Call for trained personnel to clear the scene of hazards before proceeding.

Who is responsible for determining to which facility your patient is transported?

The on-scene EMT, following off-line or online medical control procedures.

Under what circumstances might you bypass the nearest medical facility?

Trauma; burns; victim is a child; facility on bypass; specialized care.

In what two ways should you give your patient report upon arrival at the receiving facility?

Provide both an oral report and a written report.

As applied to EMS, what does "quality improvement" mean?

A system of internal/external reviews and audits of all aspects of an EMS system to identify those aspects needing improvement; assures that the public receives the highest quality of prehospital care.

Through what means does the EMT-B maintain up-to-date knowledge and skills?

Continuing education and refresher courses.

What are the six aspects of EMS quality improvement?

Documentation; run reviews and audits; gathering feedback from patients and hospital staff; continuing education; continuing preventative maintenance; skill maintenance.

As an EMT-B, you serve as the designated agent of whom?

The physician medical director of your EMS system.

Who is responsible for reviewing EMT-B quality improvement activities?

The medical director.

What level(s) of EMS provider services must have a medical director?

EMT-B, EMT-I, and paramedic.

What constitutes online medical control?

Physician medical direction by phone or radio.

What constitutes off-line medical control?

Protocols and standing orders.

What emotions may be expressed by dying patients and their families?

Rage, anger, and despair.

What does the dying patient need emotionally from the EMT?

Dignity, respect, sharing, communication, privacy, and control.

What lifestyle changes can the EMT institute to reduce stress?

Diet, exercise, and relaxation techniques.

What is a critical incident stress debriefing (CISD) team?

A team of peer counselors and mental health professionals who help emergency care workers deal with stress from difficult or unusual incidents.

What is the purpose of CISD?

To accelerate the normal recovery process after experiencing a critical incident.

What does comprehensive critical incident stress management include?

Preincident stress education; on-scene peer support; one-on-one support; disaster support services; defusings; CISD; follow-up services; spouse/family support; community outreach programs; other health and welfare programs such as wellness programs.

What types of protective masks are available to the EMT?

Surgical type for possible blood splatter and high-efficiency particulate air (HEPA) respirator if patient is suspected of or diagnosed with tuberculosis.

What is the most effective method of preventing the spread of infection from individual to individual?

Handwashing.

When is the EMT required to wear eye protection?

Any time there is a hazard of being exposed to blood or other body fluids.

When are vinyl or latex gloves needed?

When there is a possibility of contact with blood or other body fluids; they should be changed between contact with different patients.

What can help you identify possible hazardous materials spills on highways?

Binoculars; placards; *Hazardous Materials, The Emergency Response Handbook,* published by the U.S. Department of Transportation.

Who controls the hazardous materials scene?

The specialized hazardous materials team.

Fire department personnel bring you a patient from the hot zone. What is your first priority?

To determine whether the patient has been properly decontaminated.

What potential life threats are present at an accident scene?

Electricity, fire, explosion, and hazardous materials.

What articles of protective clothing should the EMT wear at a crash scene?

Turnout gear, puncture-proof gloves, helmet, and eye wear.

At a potential crime scene, the EMT must be wary that violence can be committed not only by the perpetrator, but also by the following people as well:

Family members and bystanders.

Prior to entering a potentially violent scene, what should the EMT do?

Wait for the scene to be secured by law enforcement personnel.

What immunizations should the EMT have?

Tetanus prophylaxis and the hepatitis B vaccine.

What medical tests should the EMT have done on a regular basis?

Tuberculin purified protein derivative (PPD) testing and verification of immune status with respect to commonly transmitted contagious diseases.

What are the five stages that dying patients, and those close to them, often experience?

Denial; anger; bargaining; depression; acceptance.

What are the top sources of stress for the EMT-B?

Long hours; boredom between calls; working too much and too hard; having to respond instantly; responsibility for patients' lives; making life-and-death decisions; fearing errors; dealing with the death and the grieving; receiving little recognition.

Name some high-stress situations for the EMT-B.

Multiple-casualty incidents; trauma to children; child, elder, or spousal abuse; amputations; death or injury to a coworker or other public safety worker.

What is burnout?

A state of exhaustion and irritability that can markedly decrease effectiveness.

What are the warning signs of stress?

Irritability; inability to concentrate; anxiety; difficulty sleeping; nightmares; indecisiveness; guilt; loss of appetite; loss of sexual desire; loss of interest in work; feeling of isolation.

What two things can you do to reduce stress in the work environment?

Request work shifts allowing for more time to relax with family and friends; request a rotation of duty assignment to a less busy area.

In what ways can you change your diet to help reduce stress?

Reduce sugar, caffeine, and alcohol intake; avoid fatty foods; increase carbohydrates.

Within what time frame should a CISD meeting be held?

Within 24 to 72 hours of a major incident.

What is discussed in a CISD meeting?

Feelings, fears, and reactions.

Who should evaluate the information gained and offer suggestions on overcoming stress in the CISD meeting?

CISD leaders and mental health personnel.

What is the CISD meeting designed to accomplish?

To accelerate the normal recovery process after experiencing a critical incident.

What aspects of the CISD meeting serve to accelerate the normal recovery process?

The ventilation of feelings and the nonthreatening environment.

True or false: Goggles are required for body substance isolation.

False. Goggles are NOT required.

When do you need to wear utility gloves?

When cleaning vehicles or equipment.

When do you need to wear a protective gown?

In large-splash situations such as a field delivery or major trauma.

Is wearing a gown the best way to guard against body substance contamination?

No. A change of uniform is preferred.

When should a patient wear a mask? What type?

A patient should wear a surgical mask when there is the possibility of airborne disease.

Which organizations govern the regulation of body substance isolation, notification, and testing in an exposure incident?

Occupational Safety & Health Administration (OSHA) and state agencies.

What protective clothing is used in hazardous materials incidents?

Hazardous materials suits and self-contained breathing apparatus.

What can you do to preserve a crime scene?

Do not disturb the scene unless required for medical care; help maintain the chain of evidence.

Is the EMT-B's scope of practice defined by federal, state, or local legislation?

State.

In what ways is the EMT-B's scope of practice enhanced by the medical director?

Through the use of protocols and standing orders.

To whom does the EMT have legal duties?

To the patient, medical director, and public.

What is the primary reference used by states in developing scope of practice legislation for the EMT?

The National Standard Curriculum.

Upon what three factors is the EMT-B's legal right to function contingent?

Telephone and radio communication; approved standing orders and protocols; responsibility to medical direction.

What are five basic ethical responsibilities of the EMT?

Make the physical and emotional needs of the patient a priority; practice and maintain skills to the point of mastery; attend continuing education and refresher programs; critically review performance, seeking ways to improve response time; honesty in reporting.

Give an example of an advance directive.

Do Not Resuscitate (DNR) orders.

True or false: The patient does not have the right to refuse resuscitative orders.

False.

True or false: DNR orders require a written order from a physician.

True. In most jurisdictions, DNR orders require a written order from a physician.

You arrive at the home of a 68-year-old man who presents with a weak, thready pulse and respirations of eight breaths/min. The wife states she has DNR orders, but she can't find them. What do you do?

Begin resuscitative efforts.

In order to obtain expressed consent from a patient, that patient must:

Be of legal age, be able to make rational decisions, and be informed of the steps of the procedures and all related risks.

When must you obtain expressed consent from a patient?

Before rendering treatment to every conscious, mentally competent adult.

What is implied consent?

Consent assumed from the unconscious patient requiring emergency intervention.

Upon what assumption is implied consent based?

That the unconscious patient would consent to lifesaving interventions were he or she conscious.

When does the principle of implied consent apply to children?

When life-threatening situations exist and the parent or legal guardian is not available to give consent.

When does the principle of implied consent apply to mentally incompetent adults?

When life-threatening situations exist and the legal guardian is not available to give consent.

What local issues may affect consent for treating children and mentally incompetent adults?

Emancipation issues and state regulations regarding the age of minors.

What is the legal term for unlawfully touching a patient without his or her consent?

Battery.

What is the legal term for providing emergency care when the patient does not consent to the treatment?

Assault.

True or false: A patient has the right to refuse treatment, even when that treatment may prove to be lifesaving.

True.

You arrive at the home of a 46-year-old diabetic who is semiconscious. You administer a tube of glucose. The patient regains full consciousness and does not want to go to the hospital. What do you do?

Be sure that the patient has been informed of and fully understands all the risks and consequences associated with refusal of treatment and/or transport; have the patient sign a release-from-liability form.

Who can refuse treatment or transport?

Any mentally competent adult; in the case of a child or mentally incompetent adult, the parent or legal guardian.

If you have any doubt as to whether you should or should not provide care to a patient, what should you do?

Err in favor of providing care.

How can you protect yourself from the legal consequences of patient refusal?

Ensure that you document fully and accurately.

True or false: Before leaving the scene of a patient who refuses transport, you should try to persuade the patient to be transported.

True. Always err on the side of treatment.

You suspect that your patient is drunk, but he signs the release form and tells you to leave. Are you legally liable if something happens to this patient? Why?

Yes. A patient under the influence of alcohol is not considered competent to refuse treatment.

You arrive at the home of a 25-year-old man who has sawed off his hand with a hacksaw and is refusing transport to the hospital. What should you do?

Seek the assistance of law enforcement personnel, and contact medical control.

If you are in doubt regarding how to proceed with a patient who is refusing treatment, how should you proceed?

Contact medical control and law enforcement personnel.

True or false: The EMT-B should never make an independent decision not to transport.

True.

Define abandonment.

The termination of care of a patient without assuring the continuation of care at the same level or higher.

You bring a patient into the emergency room, place him on a stretcher in the hallway, inform the unit clerk, and then leave to respond to an urgent call. The patient is later found dead where you left him. For what will you be sued?

Abandonment.

Define negligence.

The deviation from the accepted standard of care resulting in further injury to the patient.

What are the four components of negligence?

Duty to act; breach of that duty; injury or damage was inflicted, either physical or psychological; the actions of the EMT-B caused the injury or damage.

True or false: In order for there to be a duty to act, a contractual or legal obligation must exist.

True.

A patient calls for an ambulance and the dispatcher confirms that an ambulance will be sent. Do the EMTs on the ambulance have an implied or formal duty to act?

Implied.

EMTs begin treatment of a patient. Is continuing treatment an implied or formal duty to act?

Implied.

An ambulance service has a written contract with a municipality. Is this an example of an implied or formal duty to act?

Formal.

An ambulance service has a written contract with a municipality. Under what circumstances may service be refused to a patient?

If there are specific clauses in the contract that indicate when service can be refused to a patient.

Other than the legal duty to act, what other obligations does the EMT have to provide care?

Ethical and moral considerations.

When driving the ambulance not in the company's service area, an EMT observes an accident. What should he or she consider before acting?

The moral and ethical duty to act; risk management considerations; proper documentation; specific state regulations regarding duty to act; Good Samaritan regulations.

What information is considered confidential in the EMT–patient relationship?

Patient history gained through interview, assessment findings, and treatment rendered.

What is required before you may release confidential information?

A written release signed by the patient.

What must be established before you can accept a release form not from the patient?

Legal guardianship.

Under what four circumstances is a release form not required before releasing confidential patient information?

Other health care providers need to know information to continue care; state law requires reporting incidents such as rape, abuse, or gunshot wounds; third-party billing forms; legal subpoena.

Which documents are considered to provide legal permission for organ donation?

A signed intent to be a donor on the reverse of a patient's driver's license, or a signed separate donor's card.

True or false: You should treat a potential organ donor differently from any other patient requesting treatment.

False.

What is the EMT's role in organ harvesting?

Identify the patient as a potential donor; establish communication with medical direction; provide care to maintain viable organs.

In what three forms are medical identification insignia most commonly found?

Bracelets, necklaces, and cards.

What medical conditions are indicated on medical identification insignia?

Allergies, diabetes, epilepsy, and others.

What is the EMT's priority at a crime scene?

Care of the patient.

What things may the EMT do to help preserve evidence?

Do not disturb any item at the scene unless emergency care requires it; observe and document anything unusual; if possible, do not cut through holes in clothing from gunshot wounds or stabbings.

How are special legal reporting situations established?

Established by state legislation; may vary from state to state.

What are some commonly required legal reporting situations?

Child, elder, or spousal abuse; wounds obtained by violent crime; sexual assault; infectious disease exposure; patient restraint; mental incompetence.

What is normal anatomic position?

A person standing, facing forward, with the palms facing forward.

What is the midline plane?

An imaginary line drawn vertically through the middle of the body from the nose to the umbilicus; it divides the body into right and left halves.

What is the midaxillary line?

A line drawn vertically from the middle of the axilla to the ankle; it divides the body into anterior and posterior halves.

What is the torso?

The trunk of the body from the base of the neck to the top of the legs.

Define medial.

Toward the midline.

Define lateral.

Away from the midline.

Define proximal.

Nearer to a point of reference or attachment, usually the trunk of the body, than to other parts of the body.

Define distal.

Away from or the farthest from a point of reference or attachment.

Define superior.

Above or higher.

Define inferior.

Below or lower.

Define anterior.

The front or toward the front.

Define posterior.

The back or toward the back.

What is the midclavicular line?

An imaginary line that extends downward over the trunk from the midpoint of the clavicle; it divides each half of the anterior chest into two parts.

What is bilaterally?

On both the right and the left sides.

Define dorsal.

Pertaining to the back or posterior.

Define ventral.

Pertaining to the front or anterior.

Define plantar.

Refers to the sole of the foot.

Define palmar.

Refers to the palm of the hand.

Define prone.

Lying on the stomach.

Define supine.

Lying on the back.

What is the Fowler's position?

Lying on the back with the upper body elevated 45–60 degrees.

What is the Trendelenberg position?

Lying on the back with the lower body elevated approximately 12 inches; sometimes referred to as the shock position.

What is the shock position?

Lying on the back with the lower body elevated approximately 12 inches; also known as the Trendelenberg position.

What are the three functions of the skeletal system?

Gives the body shape; protects the vital internal organs; provides for body movement.

What is the function of the skull?

Houses and protects the brain.

Name the bones of the face.

Orbit; nasal bone; maxilla; mandible; zygomatic bones.

What is the mandible?

The jawbone.

What are the zygomatic bones?

The cheek bones.

What are the five sections of the spinal column?

Cervical; thoracic; lumbar; sacrum; coccyx.

How many cervical vertebrae are there?

Seven.

How many thoracic vertebrae are there?

Twelve.

How many lumbar vertebrae are there?

Five.

How many sacral vertebrae are there?

Five (fused in adults).

How many coccigeal vertebrae are there?

Four (fused in adults).

To what does the word cervical refer?

The neck.

To what does the word thorax refer?

The upper back.

To what does the word lumbar refer?

The lower back.

To what does the word sacrum refer?

The back wall of the pelvis.

To what does the word coccyx refer?

The tailbone.

How many ribs are there?

Twelve.

To what are the ribs attached posteriorly?

The thoracic vertebrae.

Which ribs are attached to the sternum?

Pairs 1–10.

To what are ribs 1–10 attached anteriorly?

The sternum.

Which ribs are floating?

Pairs 11 and 12.

To what does the term "floating ribs" refer?
Rib pairs 11 and 12, which are not attached anteriorly.

What is the sternum?
The breastbone.

What three bones make up the sternum?
The manubrium; body; xiphoid process.

What is the manubrium?
The superior portion of the sternum.

What is the body of the sternum?
The middle bone of the sternum.

What is the xyphoid process?
The inferior portion of the sternum.

What are the three parts of the pelvic girdle?
The iliac crest; the pubis; the ischium.

What is the iliac crest?
The wings of the pelvis.

What is the pubis?
The anterior portion of the pelvis.

What is the ischium?
The inferior portion of the pelvis.

Which two structures articulate at the hip joint?
The greater trochanter and the acetabulum.

What is the acetabulum?
The hip socket.

What is the greater trochanter?
The ball of the hip joint.

What is the femur?
The thigh bone.

What is the patella?
The knee cap.

What two bones comprise the lower leg?
The tibia and fibula.

What is the tibia?
The shin bone.

What are the medial and the lateral malleolus?
The surface landmarks of the ankle joint.

What bones make up the foot?
The tarsals and metatarsals.

What is the calcaneus?
The heel bone.

What bones make up the toes?
The phalanges.

What is the collarbone?
The clavicle.

What is the shoulder blade?

The scapula.

What is the point of the shoulder?

The acromion.

What is the upper arm bone?

The humerus.

What is the bone of the lateral forearm?

The radius.

What is the bone of the medial forearm?

The ulna.

What are the bones of the wrist?

The carpals.

What are the bones of the hand?

The metacarpals.

What are the bones of the fingers?

The phalanges.

What types of joints are there?

Ball and socket and hinged.

The pharynx is composed of what two structures?

The oropharynx and the nasopharynx.

What is the function of the epiglottis?

To prevent food and liquid from entering the trachea during swallowing.

What is the windpipe called?

The trachea.

What is the cricoid cartilage and where is it located?

A firm cartilaginous ring; forms the lower portion of the larynx.

What is the voice box called?

The larynx.

What are the bronchi?

The two major branches of the trachea to the lungs; each subdivides into smaller air passages called the bronchioles, which end at the alveoli.

What muscles contract during inhalation?

The diaphragm and the intercostal muscles.

What are the three basic steps of alveolar/capillary exchange?

Oxygen-rich air enters the alveoli during each inspiration; oxygen-poor blood in the capillaries passes into the walls of the alveoli; oxygen enters the capillaries as carbon dioxide enters the alveoli.

What are the two basic components of the capillary/cellular exchange in the alveoli?

Cells give up carbon dioxide to the capillaries; capillaries give up oxygen to the cells.

What is the normal breathing rate for an adult?

12–20 breaths per minute.

What is the normal breathing rate for a child?

15–30 breaths per minute.

What is the normal breathing rate for an infant?

25–50 breaths per minute.

Which two words are used to describe basic breathing rhythm?

Regular and irregular.

How would you describe normal breath sounds?

Present and equal.

How would you describe normal chest expansion?

Adequate and equal.

How would you describe increased effort of breathing?

Use of accessory muscles, predominantly in infants and children.

How would you describe normal depth of breathing?

Adequate.

You are treating a patient who is tachypnic, with shallow, inadequate respirations. What might you expect to discover when examining his skin?

The skin may be pale or cyanotic and cool and clammy.

You are called to the home of a 3-year-old girl. The parents state it has been increasingly difficult for her to breathe the past few hours, and she has used her inhaler five times. What might you expect to see on inspection of her chest and belly?

There may be retractions above the clavicles, between the ribs and below the rib cage; "seesaw" breathing may be present (the abdomen and chest move in opposite directions).

In reference to the previous case, what might you expect to see when inspecting her face?

Nasal flaring may be present; mucous membranes may be pale or cyanotic; skin may be cool and diaphoretic; the patient may appear anxious.

Why is it easier for a child's airway to become obstructed than an adult's?

All structures are smaller and more easily obstructed.

How do the adult and child pharynxes differ?

Infants and children's tongues take up proportionally more space in the mouth than adults' do.

How do the adult and child tracheas differ?

Infants and children have narrower tracheas that are obstructed more easily by swelling; the trachea is softer and more flexible; the cricoid cartilage is less developed and less rigid; it is the narrowest portion of the child's airway.

What is the principal difference in the way children and adults breathe?

The chest wall of the child is softer and the muscles less well developed; therefore, infants and children rely more heavily on the diaphragm for breathing than adults do.

What are the four chambers of the heart?

Right atrium; left atrium; right ventricle; left ventricle.

What is the function of the heart valves?

To prevent the backflow of blood.

What does the right atrium do?

Receives blood from the veins of the body and the heart; pumps the deoxygenated blood to the right ventricle.

What does the right ventricle do?

Pumps deoxygenated blood to the lungs.

What does the left atrium do?

Receives oxygenated blood from the lungs via the pulmonary veins; pumps it to the left ventricle.

What does the left ventricle do?

Receives oxygenated blood from the left atrium; pumps it to the body.

What is unique about cardiac muscle?

It is made up of special contractile and conductive tissue.

What is the function of the arteries?

To carry oxygenated blood from the heart to the rest of the body.

What are the arteries of the heart called?

The coronary arteries.

What is the function of the coronary arteries?

To supply the heart with blood.

What is the major artery originating from the heart?

The aorta.

Trace the route taken by the aorta.

The aorta originates in the left ventricle; descends anterior to the spine into the thoracic and abdominal cavities; divides at the level of the umbilicus into the iliac arteries.

What is the origin and function of the pulmonary artery?

Originates in the right ventricle; carries deoxygenated blood to the lungs.

Upon arrival at a motor vehicle crash, you find a woman in her 20s supine on the street. How do you initially assess circulation?

Palpate the carotid artery.

What is the function of the carotid arteries?

To carry oxygenated blood to the head.

What is the major artery of the thigh?

The femoral artery.

Where can you palpate the femoral artery?

In the crease between the abdomen and the thigh.

Which vessel supplies the lower arm?

The radial artery.

Where can you palpate the radial artery?

Pulsations can be palpated at the wrist thumbside.

Which artery supplies the upper arm?

The brachial artery.

Where can you palpate the brachial artery?

Pulsations can be palpated on the inside of the arm between the elbow and the shoulder.

What is another name for a blood pressure cuff?

Sphygmomanometer.

Which vessel is auscultated when taking a blood pressure?

The brachial artery.

What is the major vessel supplying the foot?

The posterior tibial artery.

Where can the posterior tibial artery be palpated?

On the posterior surface of the medial malleolus.

After the application of a traction splint, how do you evaluate distal circulation and where?

Palpate the dorsalis pedis on the anterior surface of the foot; check capillary refill time, sensation, skin condition, and movement.

What is the smallest branch of an artery?

An arteriole.

What structure connects arterioles and venules?
Capillaries.

You arrive on scene at a restaurant kitchen where you have been called for a man bleeding. When you arrive, you find a conscious, alert, and oriented 45-year-old man who states he cut his hand with a knife. You observe dark red blood oozing slowly from a 5-centimeter laceration on the palm of his left hand. From what type of blood vessel(s) do you suspect is the source of the bleeding?
Capillaries.

Where are capillaries found?
In all parts of the body.

What is the function of the capillaries?
They allow for the exchange of nutrients and waste at the cellular level.

What is the smallest branch of the venous system?
The venules.

What is the function of the venous system?
To carry deoxygenated blood back to the heart and lungs.

What is the function of the pulmonary vein?
To carry blood from the lungs to the left atrium.

Why may the pulmonary vein be considered an artery?
Because it carries oxygenated rather than deoxygenated blood.

What are largest vessels in the venous system?
The vena cava.

Name the two portions of the vena cava.
The superior and the inferior venae cavae.

What is the terminus of the vena cava?
The right atrium.

What are the four components of blood?
Red blood cells; white blood cells; plasma; platelets.

What are the two primary functions of the red blood cells?
To carry oxygen to the organs; to carry carbon dioxide from the organs.

What is the function of the white blood cells?
They function as part of the body's defense against infection.

What is another name for white blood cells?
Leukocytes.

What component of blood is the fluid that carries the blood cells and nutrients?
Plasma.

To which mechanism are platelets essential?
Clotting.

What is an erythrocyte?
A red blood cell.

Which component of the red blood cell is responsible for oxygen and carbon dioxide transport?
Hemoglobin.

What causes a pulse?
Contraction of the left ventricle, which sends a wave of blood through the arteries.

Where can a pulse be palpated?

Anywhere an artery simultaneously passes near the skin surface and over a bone.

What are the four primary sites for palpating a peripheral pulse on an adult?

The radial arteries; brachial arteries; posterior tibial arteries; dorsalis pedis arteries.

What are the two primary sites for palpating a central pulse?

The carotid and femoral arteries.

What causes systolic blood pressure?

The pressure exerted against the walls of the artery when the left ventricle contracts.

What causes diastolic blood pressure?

The pressure exerted against the walls of the artery when the left ventricle is at rest.

You are dispatched to the home of a 26-year-old woman who has slashed her wrists. Upon arrival, you find her barely conscious, sitting on the kitchen floor. There are at least two liters of fresh red blood on the floor. You lean down to speak with her as your partner, after donning gloves, gown, and mask, applies direct pressure with trauma pads to her lacerations. What do you notice about her skin?

It is pale, cyanotic, cool, and clammy.

You reach down to palpate the above patient's pulse. You are unable to palpate her radial pulse due to the wounds. Your next choice would be where?

The brachial artery.

What rate and quality of pulse do you expect for the wounds listed above?

Rapid, weak, and thready.

You auscultate the wounded 26-year-old's chest to assess the rate and quality of breathing. What do you expect to find?

Rapid, shallow breathing.

The patient won't sit still and weakly tries to get up. She states she has to get out of the house and that she is sick to her stomach. Are these appropriate behaviors given the patient's condition?

Yes. Restlessness, anxiety or mental dullness, nausea, and vomiting are all symptoms of shock.

You auscultate a blood pressure of 80/60 mm Hg and take an axillary temperature of 34°C. Are these normal findings given the patient's condition?

Low or decreasing blood pressure (hypotension) and subnormal temperature are both signs of late, decompensated shock.

What is the definition of perfusion?

The circulation of blood through an organ or a structure.

What is the function of perfusion?

The delivery of oxygen and other nutrients to the cells of all organ systems and the removal of waste products.

What is hypoperfusion?

The inadequate circulation of blood through an organ or a structure.

What are the three types of muscle?

Voluntary (skeletal); involuntary (smooth); cardiac.

To what do skeletal muscles attach?

Bones.

Which muscles are responsible for movement?

Skeletal.

Which muscles are responsible for carrying out automatic muscular functions?

Smooth.

Which muscles are under the control of the brain and nervous system?

Skeletal.

Which kind of muscle controls the blood vessels?

Smooth.

Which kind of muscle is found in the urinary and intestinal tracts?

Smooth.

Which kind of muscle possesses the quality of automaticity?

Cardiac.

In what way are cardiac and smooth muscle similar?

They are both involuntary.

What is the function of the nervous system?

To control the voluntary and involuntary activity of the body.

What are the two components of the nervous system?

The central and peripheral.

What are the two components of the central nervous system?

The brain and spinal cord.

What are the two functions of the peripheral nervous system?

Sensory and motor.

What are the functions of the skin?

Protection, temperature regulation, and sensation.

What are the three main layers of the skin?

The epidermis; dermis; subcutaneous layer.

Which structures are contained within the dermis?

Sweat and sebaceous glands; hair follicles; blood vessels; nerve endings.

What is the function of the endocrine system?

To secrete chemicals, such as insulin and adrenaline, which are responsible for regulating body activities and functions.

What general information do you obtain from a patient?

Chief complaint, age, sex, and race.

What is the chief complaint?

The reason EMS was notified.

What baseline vital signs should you obtain?

Breathing, pulse, perfusion, pupils, and blood pressure.

When assessing breathing, what two main values are you looking for?

Rate and quality.

What is the recommended method for calculating the rate of breathing?

Count the number of breaths in a 30-second period, then multiply by two; to avoid influencing the rate, care should be taken not to inform the patient.

Quality of breathing can be placed in one of four categories. What are they?

Normal; shallow; labored; noisy.

What are some of the characteristics of labored breathing?

Increased effort of breathing; grunting and stridor; use of accessory muscles; nasal flaring; supraclavicular and intercostal retractions in infants and children; sometimes gasping.

What may be included in noisy respirations?

Snoring, wheezing, gurgling, and crowing.

You are called to the home of a 5-year-old with a history of asthma. The patient is tripoding and retracting. What sound do you expect to hear on auscultation?

Wheezing. This child is showing the signs of an acute asthma attack; as the attack progresses and the airways become increasingly constricted and occluded, you may have difficulty auscultating breath sounds at all.

You arrive at a nursing home to transport a 76-year-old overweight man. He is sitting upright, is tachypnic and working hard to breathe, and his ankles are swollen. What do you expect to hear on auscultation?

Gurgling. This patient is showing the signs of congestive heart failure, which causes fluid to back up into the lungs.

You are transporting a 6-month-old girl with a fever. Where would you palpate her pulse?

Over the brachial artery.

A mother walks into your station carrying her 2-year-old son. She states his heart is "racing." Where would you palpate his pulse?

Over the radial artery.

When palpating a pulse, what two things are you looking for?

Rate and quality.

What is the recommended method for measuring a pulse rate?

Count for 30 seconds, then multiply by two.

In what four ways can you characterize the quality of a pulse?

Strong; weak; regular; irregular.

Your patient at the nursing home with CHF collapses on his bed while you are conducting your assessment. Your partner, and the nurse who is helping from the other side of the bed, palpate the patient's carotid pulse bilaterally. What do you do?

Ask one of them to stop; bilateral carotid compression can compromise cerebral circulation, as well as increase the risk of an unwanted vagal response.

Your partner removes his hand, but the nurse continues to try to find the carotid pulse. She rubs her fingers back and forth across the patient's neck, desperately searching for a pulse. What is she doing wrong?

Excess carotid pressure on geriatric patients can cause the dislodging of clots, which form emboli that can result in cerebral vascular accident (stroke).

Where should you assess skin color?

Nail beds, oral mucosa, and conjunctiva.

In addition to the previous noted sites, where is a good place to assess skin color on a child or infant?

Palms of hands and soles of feet.

What does pale skin indicate?

Poor perfusion and impaired blood flow.

What is cyanosis?

A blue-gray color.

What does cyanosis indicate?

Inadequate oxygen or poor perfusion.

What does flushed skin indicate?

Exposure to heat or carbon monoxide poisoning.

What does jaundice indicate?

Liver abnormalities.

You are called to the home of an elderly woman who complains of weakness. She appears pale and confused. You feel her skin with the back of your hand, and it is cool to the touch. What does this indicate?

Poor perfusion or exposure to cold.

You are evaluating a 40-year-old man complaining of shortness of breath. He states he has been sitting quietly and reading. When you palpate his skin, it is wet and cool. Is this a normal finding?

No. The temperature indicates inadequate perfusion; the wetness is clearly an abnormal response in a resting patient.

Under what age is the capillary refill test indicated?

Less than 6 years.

What is considered abnormal capillary refill time in infants and children?

Greater than 2 seconds.

What is important to look for in assessing the pupils?

Reactivity and equality.

What are the two methods the EMT can use in measuring a blood pressure?

Palpation and auscultation.

Above what age should a blood pressure be measured in all patients?

Greater than 3 years.

You arrive at a day-care center to transport a 2-year-old girl to the emergency department for evaluation. The patient is pale and difficult to arouse, and her skin is cool and damp to the touch. Her vital signs HR 120, RR 34, CRT <2, temp 37. The nurse at the day-care center tells you she is just sleepy, her vital signs are normal, and the teacher is overreacting. What do you do?

Explain to the nurse that even sick children may sustain normal vital signs and that the child looks ill and should be transported for evaluation; the general assessment of the infant or child, such as appearing sick, in respiratory distress, or unresponsive, is more valuable than vital sign numbers.

En route to the hospital, how often should you reassess this patient's vital signs?

Every 5 minutes; assess and record vital signs every 5 minutes in an unstable patient.

You decide to administer oxygen. Should you reassess vital signs?

Yes. Vital signs should be reassessed after every intervention.

You are transporting a stable, 65-year-old man to the hospital for an X-ray of his broken leg prior to having the cast removed. How often should you take his vital signs en route?

Vital signs should be assessed and recorded every 15 minutes at a minimum in a stable patient.

What is a SAMPLE history?

Signs/Symptoms; Allergies; Medications; Pertinent medical history; Last oral intake; Events leading to the injury or illness.

You are evaluating a 24-year-old woman for chest pain. She states she is short of breath. Is this a sign or a symptom?

A symptom, which is any condition described by the patient.

You note the woman is retracting and tachypnic. Are these signs or symptoms?

Signs, which are anything observable by the EMT.

What kinds of allergies are important to ask about?

Medications, food, or environmental.

How might you go about eliciting allergy information in an unconscious patient?

Look for a medic alert necklace, bracelet, or wallet card.

What kind of medications are important to find out about?

All recent or current prescription and nonprescription medications, including birth control pills; consider looking for a medic alert bracelet, necklace, or wallet card.

What kind of pertinent past history is it important to find out about?

Medical, surgical, or trauma.

What information are you trying to elicit in regard to last oral intake?

Solid or liquid; time and quantity.

You are evaluating a 42-year-old man who is complaining of chest pain. What is important to know regarding the events that led up to his experiencing this pain?

What was he doing? Was he resting or exerting himself?

What are the two most common errors committed when lifting patients?

Using the back to lift, and not keeping weight as close to the body as possible.

You and your partner arrive on the third floor of a walk-up apartment house to find a very large woman stuck in the bathtub. She is flowing over the sides of the tub. She is in no distress. What should you do?

Consider the need for additional help.

What self-assessment do you need to make before attempting to lift a patient?

An honest assessment of your own physical ability and limitations.

With his feet planted, your partner reaches behind him to pick up the airway kit. What did he do wrong?

He twisted his body while lifting instead of turning to align himself with the object; he did not have his feet positioned properly for the lift.

You have been working with a new partner for 2 weeks, and you always come home with a sore back after your shift. What's the problem?

You need to communicate more clearly and frequently with your partner to synchronize your movements.

An elderly man is to be transported to the emergency department for evaluation of a fever of several days' duration. He is dressed and sitting in a chair in his bedroom on the second floor and states he cannot walk. How do you get him downstairs?

With a stair chair.

You and your partner are about to lift a patient whom you have secured to a backboard onto your stretcher, when three fire fighters arrive to help. Should all of you pitch in?

No. Use an even number of people to lift so that balance is maintained.

After removing a very large woman from the bathtub with the assistance of the fire department, you place her on your stretcher, which promptly collapses. What did you do wrong?

You did not know the weight limitations of your equipment.

You are at a loss now and don't know what to do. How could you have prepared for this situation?

Know what to do with patients who exceed weight limitations of equipment.

What two techniques can you use to help improve your lifting power and safety?

The power lift and the power grip.

When preparing to carry a patient, what is important to know?

The weight of the patient and the crew's abilities and limitations.

Is it better to flex your hips and knees or your waist when carrying?

Hips and knees.

True or false: You should always lean back from the waist when carrying.

False. Do not hyperextend the back.

True or false: Keep your back in a locked-in position when carrying.

True.

True or false: It is not important for partners to have similar strength and height when carrying.

False.

True or false: When possible, use a stretcher instead of a stair chair when carrying a patient.

False.

When carrying, lifting, and reaching, what type of movement is it important to avoid?

Twisting.

When reaching, it is important to avoid reaching how many inches in front of the body?

15–20.

True or false: Situations where prolonged (more than a minute) strenuous effort is needed can cause injury.

True.

Describe the correct reaching technique for log rolls.

Keep the back straight while leaning over the patient, lean from the hips, and use shoulder muscles whenever possible.

True or false: Pull, rather than push, whenever possible.

False.

When pulling, how can you keep the line of pull through the center of your body?

Bend your knees.

True or false: Unlike in lifting, when pulling, it is important to keep the weight away from the body.

False.

What is the best area of your body to push from?

The area between the waist and the shoulder.

If the weight being pushed is below waist level, what should you do?

Use the kneeling position.

True or false: When pushing, you should keep your arms straight and extend them from the shoulder.

False. Keep the elbows bent with the arms close to the sides.

Under what circumstances is it important to move a patient immediately?

When there is danger of fire, explosives, and other hazardous materials; when you cannot protect the patient from scene hazard; when you cannot gain access to other patients in a vehicle who need lifesaving care; when the patient's position precludes you from providing appropriate care.

Under what circumstances is it important to move a patient quickly?

Altered mental status, inadequate breathing, or shock.

What is the greatest danger in moving a patient quickly?

Aggravating a spinal injury.

What can you do to limit the danger of aggravating a spinal injury?

Make every effort to pull the patient in the direction of the long axis of the body; this provides as much protection to the spine as possible.

While treating an unconscious patient at the scene of a water heater explosion, the fire fighters tell you to evacuate the house because of the danger of a flashover. What three techniques can you use to rapidly extricate the patient?

Pull on the patient's clothing in the neck or shoulder area; put the patient on a blanket and drag the blanket; put your hands under the patient's armpits from the back or grasp the patient's forearms and drag.

Your crew arrives at the scene of a motor vehicle crash. A four-door sedan is upright, with its front end wrapped around a utility pole. The middle-aged man in the driver's seat is unresponsive, pulse 130, weak and thready, respirations 8 and gasping. You are the crew chief, and have Sam, Jacob, and Cheri on your truck. What do you tell them to do?

Sam: Get behind the patient and provide manual immobilization.
Jacob: Apply cervical immobilization device.
Cheri: Place the long backboard near the door, then move to the passenger seat.
All: On Jacob's count, rotate the patient in several short, coordinated moves until his back is in the open doorway and his feet are on the passenger seat.
Sam: Pass manual immobilization to someone else, get out of the car, and reassume.
All: Place the end of the long backboard on the seat next to the patient's buttocks; support the other end of the board; Sam and Jacob lower the patient to the board and slide him into the proper position with short, coordinated moves.

When lifting a patient from the ground with no risk of spinal injury, do the rescuers stand on the same or opposite sides of the patient?

On the same side.

True or false: When using an extremity lift, the EMT at the head grasps the wrists.

True.

When pulling a stretcher, which end should you grasp?

The foot end.

When loading an ambulance, should hanging or wheeled stretchers be loaded first?

Hanging stretchers.

What two types of backboards are there?

Traditional wooden and manufactured.

What two types of short backboards are there?

Traditional wooden and vest-type device.

You arrive on scene at a nursing home to find an unresponsive elderly woman supine in bed. How should you position her?

Patient should be rolled to the recovery position, onto the side (preferably the left), without twisting the body.

You are treating a patient who has called for an ambulance because of shortness of breath with chest pain. How should you position him or her on the stretcher?

In a position of comfort, as long as hypotension does not exist.

You are transporting a patient from a sporting event who has collapsed due to dehydration. How should he or she be positioned?

With legs elevated 8–12 inches.

What is an early intervention for a pregnant patient with hypotension?

Position her left lateral recumbent (lying on her left side).

How do you position a patient who is nauseated or vomiting?

In a position of comfort, as long as airway can be managed by the EMT.

Preparatory

1. As an EMT-B, your primary responsibilities are:
   A. Personal safety.
   B. Safety of everyone at the scene.
   C. Safety of others.
   D. Both A and C are correct.

2. The purpose of a focused history and physical exam is to:
   A. Identify and treat life-threatening problems.
   B. Obtain diagnostic information.
   C. Obtain ABCs.
   D. All of the above

3. Scene safety means your:
   A. Responsibility during transport.
   B. First responsibility when arriving on a scene.
   C. Responsibility to victims and bystanders.
   D. Responsibility to all rescuers.

4. At a motor vehicle crash with downed utility wires around the vehicle, your first responsibility is to:
   A. Get all victims safely out of the vehicle.
   B. Control traffic at or near the scene.
   C. Call for back-up.
   D. Call for trained personnel to clear the scene of hazards before proceeding.

5. As an EMT-B, you serve as the designated agent of whom or what?
   A. The physician director of the office of emergency medical training
   B. The physician medical director of your EMS system
   C. The physician in charge of the emergency room
   D. The chief medical examiner

6. Protocols and standing orders constitute:
   A. Off-line medical control.
   B. Online medical control.
   C. Your agency's policies.
   D. Both online and off-line medical control.

7. As an EMT, you have legal duties to the patient and to:
   A. His or her immediate family or survivors.
   B. Your agency and medical director.
   C. Medical director and public.
   D. Emergency room staff and public.

8. You need to wear utility gloves when:
   A. You provide all patient care.
   B. Your patient is bleeding profusely.
   C. You are cleaning your equipment or vehicles.
   D. You are assisting with extrication.

9. Duty to act, breach of that duty, physical or psychological injury or damage inflicted to the patient, and the actions of the EMT-B that caused the injury or damage to the patient are the four:
   A. Acts of omission.
   B. Components of negligence.
   C. Stages of negative care.
   D. Stages of duties to care.

10. You should treat a potential organ donor differently from any other patient requesting treatment.
A. True
B. False

11. Medical identification insignias, often called medical alert insignias, indicate:
A. Medications that the patients take.
B. That patients have medical insurance, including Medicare and Medex.
C. Allergies, diabetes, epilepsy, and other conditions.
D. Medical care requirements and religious choice (i.e., call a rabbi, call a priest).

12. Whether or not an EMT is a law-enforcement officer, at a crime scene, the EMT's first responsibility is to:
A. Call for back-up.
B. Call for police detectives.
C. Call for paramedics.
D. Care for the patient.

13. Below or lower, as it relates to the patient, is defined as:
A. Inferior.
B. Superior.
C. Posterior.
D. Distal.

14. Supine is defined as lying on the back, prone is defined as lying on the:
A. Left (heart) side.
B. Right side.
C. Stomach.
D. Back with the upper body elevated 45–60 degrees.

15. The sections of the spinal column from top to bottom have five sections; the top (first) section is cervical and the bottom (fifth) is the coccyx. The correct order of the second, third, and fourth sections is:
A. Thoracic, lumbar, and sacrum.
B. Lumbar, thoracic, and sacrum.
C. Sacrum, thoracic, and lumbar.
D. Sacrum, lumbar, and thoracic.

16. The two bones that comprise the lower leg are the tibia and fibula. Which of those are also referred to by a common name?
A. Tibia is referred to as the "tail bone."
B. Fibula is referred to as the "shin bone."
C. Tibia is referred to as the "shin bone."
D. Fibula is referred to as the "thigh bone."

17. The function of the ______________ is to prevent food and liquid from entering the trachea during swallowing.
A. Oropharynx
B. Trachea
C. Larynx
D. Epiglottis

18. How would you describe normal breath sounds?
A. Present and equal
B. Adequate and equal
C. Regular and irregular
D. Loud and soft

19. The function of the ______________ is to carry oxygenated blood from the heart to the rest of the body.
A. Vessels
B. Arteries
C. Coronary arteries
D. Capillaries

20. A sphygmomanometer is usually known as a/an:
   A. Stethoscope.
   B. AED.
   C. Blood pressure cuff.
   D. Device to measure the pulse, when attached to a finger.

21. Snoring, wheezing, gurgling, and crowing are all signs of:
   A. Noisy inhalations.
   B. Noisy exhalations.
   C. Noisy, labored perfusion.
   D. Noisy respirations.

22. When palpating a pulse, what two things are you looking for?
   A. Ease and rate
   B. Character and quality
   C. Rate and quality
   D. Noisiness and rate

23. In infants and children, what is considered abnormal capillary refill?
   A. Less than 3 seconds
   B. Greater than 2 seconds
   C. Greater than 15 seconds
   D. Less than 25 seconds

24. When you transport a stable patient to the hospital, how often should you take his or her vital signs en route?
   A. Assess and record every 5 minutes
   B. Assess and record every 10 minutes
   C. Assess and record every 15 minutes
   D. Assess at least once during transport

25. Circumstances may require you to move a patient immediately. Which of the following would not require an immediate move, but would require the patient to be moved quickly?
   A. Danger of fire, explosion, or hazardous materials
   B. Altered mental status
   C. Patient's position precludes you from providing appropriate care
   D. When you cannot protect the patient from scene hazards

Section

Airway

Voices of Experience

My First Cricothyrotomy

We were almost at the end of the shift, finishing up run reports in the emergency department, when we were dispatched to a motorcycle accident. We responded to the scene, along with the supervisor. It was dispatched as a single vehicle accident.

Perry Township Fire was already on the scene, along with two off-duty paramedics who happened to be nearby when the call went out. By the time we arrived, they had the 30-year-old male patient in a cervical collar, they were holding c-spine control, and they had the patient turned onto his left side. There was a fair amount of blood coming from the patient's mouth and nose and he was unconscious. The patient had a patent airway as long as he was in the recovery position. The paramedics gave the patient oxygen and we examined the patient; there were no other apparent injuries noted. There was a good radial pulse but it was tachy, about 120. There was a strong odor of ethanol on or about the patient.

The other paramedics and the supervisor were working on the airway, trying to get the patient intubated. Meanwhile, I obtained vital signs, started an IV of normal saline, and drew labs. We log rolled the patient onto a long backboard, checked his back, and secured him to the board. The paramedics had still not obtained an advanced airway, and the patient had clenched down so they couldn't even open his mouth to suction it. The patient was becoming very agitated and combative due to hypoxia. I said, "OK, guys, it's time to cric this guy." They all looked puzzled. The supervisor said he would call medical control to get the order, and I informed him that we didn't need one, and that the cric kit was right there in the oxygen duffle. He picked it up and tossed it to me and said, "There you go."

I wasn't even at the head of the patient. I looked at the three other medics and they all just kind of looked away. I said, "OK."

I prepped and cleaned the area of the neck and was trying to locate the proper landmark to make a small incision, which had to be made with that particular cric kit, but the larynx kept moving up and down because the patient was still gasping for air and didn't have a patent airway. I had to stick the large needle into the notch and advance the dilator into the trachea. This was rather difficult because I wasn't able to make the incision, so I had to stretch the skin more than I would have if the incision was made. I was able to advance the airway into the trachea and secure it in place. I hooked up the bag-mask device and gave the patient about three good breaths of oxygen; the patient immediately became calm and flaccid.

The fire fighters were all staring and thought that I had killed the patient. This was the first cric they had ever seen done in the field, and they didn't understand that the reason the patient was so combative was because of hypoxia. When I was able to give him the oxygen that he so desperately needed, he calmed down and relaxed.

We were then able to get the patient loaded in the ambulance and transported to the hospital. I felt pretty good for getting the cric and getting it properly placed. The emergency department doctor asked how many I had done before, and I said that it was the first cric that I had done but that I had done one needle cric before. He said, "Well, that is one more than I have ever done."

I was able to visit the patient in critical care the next day and found out that he had just bought the motorcycle, a Harley, and was celebrating with friends. He had lost control on a corner and "kissed" a few mailboxes as he left the road. He had seven fractures of the mandible and maxilla. It took four operations to get everything straightened out. The doctor was there while I was visiting, and he let me check out the placement of the cric tube through the scope. He said it was placed perfectly, and he left it in for 8 days until all the operations were done. I got to see the patient about 8 months later, and I couldn't even see a scar on his neck.

I practiced cricothyrotomies since my initial training many years ago; I practiced them at every refresher class and recertification. I had never done one before this call, nor have I met anyone who has done one since. Calls like this one are proof that on any given day, long-practiced and never-used skills can become the difference between life and death for a patient—so keep on practicing!

Dave Linton, EMT-P/Instructor
Bloomington Hospital Ambulance Service
Pellham Emergency Training, Inc.
Bloomington, Indiana

SECTION

2

Airway

"For a man to attain to an eminent degree in learning costs him time, watching, hunger, nakedness, dizziness in the head, weakness in the stomach, and other inconveniences."

—Miguel De Cervantes

Section 2 covers the following subject:

Lesson 2-1 Airway

Teaches airway anatomy and physiology, how to maintain an open airway, pulmonary resuscitation, variations for infants and children and patients with laryngectomies. The use of airways, suction equipment, oxygen equipment and delivery systems, and resuscitation devices will be discussed in this lesson.

Describe the three steps of alveolar capillary exchange.

Oxygen rich air enters the alveoli during each respiration; oxygen-poor blood in the capillaries passes into the alveoli; oxygen enters the capillaries as carbon dioxide enters the alveoli.

Describe the two steps of capillary/cellular exchange.

Cells give up carbon dioxide to the capillaries; capillaries give up oxygen to the cells.

An EMT is adequately ventilating a patient when:

The chest rises and falls with each ventilation, the rate is appropriate to the age of the patient, and the heart rate returns to normal.

Artificial ventilation is inadequate when:

The chest does not rise and fall with ventilation, the rate is too slow or too fast for the age of the patient, and the heart rate does not return to normal.

What is the most common technique for opening the airway?

The head tilt–chin lift maneuver.

What technique would you use for opening the airway when you suspect spinal injury?

The jaw-thrust maneuver.

After performing the head tilt–chin lift, you note secretions in the patient's oropharynx. What do you do?

Suction.

What is the purpose of suctioning?

To remove blood, other liquids, and food particles from the airway.

What kinds of things may suctioning prove inadequate to remove?

Teeth, foreign bodies, and food.

You are bagging a patient and hear a gurgling sound with each ventilation. What should you do?
Suction.

What two types of portable suction devices are there?
Electrical and hand-operated.

What two types of suction catheters are there?
Hard or rigid (tonsil tip) and soft (French).

How far should you insert a rigid catheter when suctioning?
Only as far as you can see.

When might you use a soft suction catheter rather than a rigid?
To suction the nasopharynx, ET tubes.

How far should you insert a soft catheter into the pharynx?
Only as far as the base of the tongue.

How much vacuum should a suction device be able to generate?
300 mm Hg.

What is the main disadvantage of a battery-operated suction unit?
Discharged batteries.

True or false: Never insert a suction catheter without suction.
False.

What is the maximum time you should apply suction to a patient?
15 seconds at a time.

In infants and children, should this time be shorter or longer?
Shorter. Children have less capacity to tolerate hypoxia.

What should you do if the patient has secretions or emesis that cannot be removed quickly and easily by suctioning?
The patient should be log rolled and the oropharynx should be cleared.

What is the procedure for handling a patient producing frothy secretions as rapidly as suctioning can remove?
Suction for 15 seconds, ventilate for 2 minutes, then suction for 15 seconds, and continue in that manner; consult medical direction for this situation.

You are suctioning a patient and the tubing becomes clogged. What should you do?
Attempt to clear the clog by suctioning water.

In order of preference, the four methods for ventilating a patient by the EMT are:
Mouth-to-mask; two-person bag-mask; flow-restricted, oxygen-powered ventilation device; one-person bag-mask.

Before beginning artificial ventilations of any kind, what should the EMT consider?
Body substance isolation.

What oxygen liter flow should you use when performing mouth-to-mask ventilations?
15 L/min.

What are the components of the bag-mask?
Self-inflating bag, one-way valve, face mask, oxygen reservoir; to perform most effectively it needs to be connected to oxygen.

What is the approximate volume of the self-inflating bag?
1,600 mL.

True or false: The bag-mask provides less volume than mouth-to-mask.
True.

True or false: The single EMT may have difficulty maintaining an airtight seal.
True.

True or false: Two EMTs using the bag-mask will be more effective than one.

True.

True or false: Position yourself at the side of the patient's head for optimal performance.

False. You should be positioned at the top of the patient's head.

What adjunctive airways may be necessary to effectively ventilate with the bag-mask?

Oropharyngeal or nasopharyngeal airways.

What characteristics should the self-refilling bag have?

It should be easy to clean and sterilize.

What kind of valve should the bag-mask have?

A nonjam valve that allows a maximum of oxygen inlet flow of 15 L/min.

Why must there be no pop-off (or a disabled pop-off) valve on the bag-mask?

It may result in inadequate ventilation.

What is the purpose of the oxygen reservoir?

It allows for a higher concentration of oxygen.

What sizes of masks should you carry for the bag-mask?

Infant, child, and adult.

Where should you position the apex of the mask of the bag-mask?

Over the patient's nose.

When using two hands to secure a mask for ventilation, which fingers hold the mask down?

The thumbs.

When using one hand to secure a mask for ventilation, which fingers hold the mask down?

The thumb and index fingers.

How often should you repeat ventilations on an adult?

Every 5 to 6 seconds.

How often should you repeat ventilations on a child?

Every 3 to 5 seconds.

If while ventilating a patient the chest does not rise and fall, what is the first thing you should do?

Reposition the head.

If while ventilating a patient the chest does not rise and fall, what should you do after repositioning the head?

If air is escaping from under the mask, reposition fingers and mask.

If while ventilating a patient the chest does not rise and fall, what should you do after repositioning fingers and mask?

Check for obstruction.

If while ventilating a patient the chest does not rise and fall, what should you do after checking for obstruction?

Use an alternative method of artificial ventilation, e.g., pocket mask, manually triggered device; if necessary, consider the use of adjuncts such as oral or nasal airways.

What precautions should you take in ventilating a patient with suspected trauma or neck injury?

Immobilize the head and neck; have an assistant immobilize or immobilize between your knees.

What type of ventilatory device is contraindicated in children?

Oxygen-powered ventilation devices.

What peak flow rate and percent of oxygen should a flow-restricted oxygen-powered ventilation device be capable of delivering?

100% at up to 40 L/min.

At what pressure should the inspiratory pressure relief valve activate on a flow-restricted oxygen-powered ventilation device?

60 cc/water.

In addition to a pressure-relief valve, what safety features should a flow-restricted oxygen-powered ventilation device have?

An audible alarm that sounds whenever the relief-valve pressure is exceeded.

How should the trigger be positioned on a flow-restricted oxygen-powered ventilation device?

In such a way that both hands of the EMT can remain on the mask to hold it in position.

What is a tracheostomy?

Permanent artificial opening in the trachea.

What special procedures do you need to use when ventilating a tracheostomy patient?

If unable to artificially ventilate, try suction, then artificial ventilation through the nose and mouth; sealing the stoma may improve ability to artificially ventilate from above or may clear obstruction. You need to seal the mouth and nose when air is escaping.

How do you use a bag-mask to stoma?

Use infant and child mask to make seal; the technique is otherwise very similar to ventilating through the mouth. The head and neck do not need to be positioned.

When may an oropharyngeal airway be used?

In assisting in maintaining an open airway on unresponsive patients without a gag reflex.

How do you select the proper size of oral airway?

Measure from the corner of the patient's lips to the tip of the earlobe or angle of the jaw.

What is the preferred method for inserting an oral airway in an adult?

Open the mouth, insert the airway upside down, advance until resistance is encountered, then turn the airway 180 degrees so that it comes to rest with the flange on the patient's teeth.

What is the preferred method of inserting an oral airway in a pediatric patient and is an alternative method in an adult?

Insert the airway right side up, use a tongue depressor to press the tongue down, and avoid obstructing the airway.

When may a nasopharyngeal airway be used?

On patients who are responsive but need assistance keeping the tongue from obstructing the airway; less noxious than the oral airway.

How do you select the proper size of nasal airway?

Measure from the tip of the nose to the tip of the patient's ear; also, consider the diameter of the airway in the nostril.

How do you insert the nasal airway?

Lubricate the airway with a water-soluble lubricant and insert it posteriorly; the bevel should be toward the base of the nostril or toward the septum.

What do you do if you are unable to advance a nasal airway?

Try the other nostril.

What is the capacity of a D cylinder?

350 L.

What is the capacity of an E cylinder?

625 L.

What is the capacity of an M cylinder?

3,000 L.

What is the capacity of a G cylinder?

5,300 L.

What is the capacity of an H cylinder?

6,900 L.

Why is it important to handle oxygen tanks carefully?

Their contents are under pressure.

What is the most delicate part of the oxygen cylinder?

The valve and gauge assembly.

What precautions can you take to avoid damaging oxygen cylinders?

Position them to prevent falling, and secure them during transport.

Approximately how many psi does a full oxygen cylinder hold?

2,000.

When is it important to humidify oxygen?

If the patient is going to be on it for an extended period.

What two steps must you take before attaching the regulator-flowmeter to the oxygen tank?

Remove the protective seal; quickly open and then shut the valve.

After using an oxygen tank, what steps should you take to secure the system?

Close the valve and bleed remaining oxygen from the regulator.

What is the preferred method of giving oxygen to prehospital patients and why?

Nonrebreathing mask; delivers up to 90% O_2.

What common error is made when placing a nonrebreathing mask on a patient?

Bag not full before placing mask.

How do you determine the correct flow rate for the nonrebreathing mask?

When the patient inhales, the bag should not collapse.

Regarding what types of patients are there unwarranted concerns about giving "too much" oxygen?

COPD patients and infants.

What does research indicate in regard to the administration of high concentrations of oxygen to COPD patients and infants in the prehospital setting?

Research has shown that concerns about the dangers of giving these patients too much oxygen are not valid; all patients who require oxygen (are cyanotic, cool, clammy, short of breath) should receive high-concentration oxygen.

What size oxygen masks should be carried on every ambulance?

Infant, pediatric, and adult.

When should a nasal cannula be used?

Only when the patient will not tolerate a nonrebreathing mask, despite coaching from the EMT.

You arrive at a playground where a 6-month-old child is being held by her frantic mother. You assess the child and find she is not breathing. In what position do you place this child's head for ventilation?

In the correct neutral position.

What is the correct head position for ventilating a child?

Extended slightly past neutral.

Why is it important not to hyperextend the head of an infant or child when ventilating?

Their trachea and necks are very pliable, and you may occlude the airway.

What special concerns do you have in using the bag-mask on a pediatric patient?

Avoid excessive bag pressure; use only enough to make the chest rise because more than that can damage the lungs and inflate the stomach.

True or false: When ventilating a child, the pop-off valve should always be activated.

False.

True or false: Gastric distention is more common in adults than in children.

False.

If you are unable to ventilate a child using the head tilt, jaw thrust, or chin lift, what should you do?

Consider an oral or nasal airway.

What problems do facial injuries pose to establishing and maintaining a patent airway?

Because the blood supply to the face is so rich, blunt injuries to the face frequently result in severe swelling; for the same reason, bleeding into the airway from facial injuries can be a challenge to manage.

True or false: Ordinary dentures should always be removed before ventilating a patient.

False.

What is the concern with partial dentures when ventilating a patient?

They may become dislodged and occlude the airway.

Airway

1. When the chest does not rise and fall with ventilation, the rate is too slow or too fast for the age of the patient, and the heart rate does not return to normal, what is indicated?
A. Inadequate artificial ventilation
B. Inadequate capillary/cellular exchange
C. Obstructed airway
D. A need to attach an AED

2. Which of the following is the most common technique for opening the airway?
A. Modified jaw-thrust maneuver
B. Head tilt–chin lift maneuver
C. Head tilt–neck lift maneuver
D. Tongue–jaw-thrust maneuver

3. What technique would you use for opening the airway when you suspect spinal injury?
A. Head tilt–chin lift maneuver
B. Tongue–jaw-thrust maneuver
C. Jaw-thrust maneuver
D. Don't move the head or neck.

4. Which of the following things may not be removed during suctioning?
A. Blood
B. Liquids other than blood
C. Food particles
D. Teeth

5. How far should you insert a soft catheter into the pharynx?
A. To the tonsils, adenoids
B. Base of the tongue
C. 3 inches past the tongue
D. Only as far as you can see

6. How much vacuum should a suction device be able to generate?
A. 100 mm Hg
B. 200 mm Hg
C. 300 mm Hg
D. 500 mm Hg

7. Place the following four methods for ventilating in order of preference.
1. One-person bag-mask
2. Mouth-to-mouth mask
3. Flow-restricted, oxygen-powered ventilation device
4. Two-person bag-valve-mask

A. 2, 1, 3, 4
B. 1, 2, 3, 4
C. 2, 4, 3, 1
D. 4, 3, 2, 1

8. What oxygen-liter flow should you use when performing mouth-to-mouth mask ventilations?
A. 15 L/min
B. 18 L/min
C. 20 L/min
D. 20–25 L/min

9. What is the maximum time you should apply suction to a patient?
   A. 10 seconds at a time
   B. 15 seconds at a time
   C. 30 seconds at a time
   D. No greater than 1 minute at a time
10. Based on your EMT training, before you begin artificial ventilation of any kind, what should you consider?
   A. Age and weight of the patient
   B. Contagious disease properties
   C. Body substance isolation
   D. Location of patient
11. The bag-mask device provides less volume than mouth-to-mask.
   A. True
   B. False
12. What sizes of masks should you carry for the bag-mask?
   A. Adult only
   B. Infant and child
   C. Adult and child
   D. Infant, child, and adult
13. If, while ventilating a patient, the chest does not rise and fall, what is the first thing you should do?
   A. Reposition the head.
   B. Lift the chin higher.
   C. Ventilate with greater ventilation volume.
   D. Change the method you are using to ventilate.
14. When an EMT uses a bag-mask for a stoma, use an infant or child mask.
   A. True
   B. False
15. The head and neck do not need to be positioned when using a bag-mask on a stoma patient.
   A. True
   B. False
16. What should you do if you are unable to advance a nasal airway?
   A. Relubricate the airway and try again.
   B. Try the other nostril.
   C. Stop attempting, so you don't injure patient.
   D. All of the above
17. The capacity of an E cylinder oxygen tank is 625 L. What size cylinder holds less capacity (350 L)?
   A. A
   B. B
   C. C
   D. D
18. After using an oxygen tank, what steps should you take to secure the system?
   A. Close the valve.
   B. Disinfect the attachments.
   C. Close the valve and bleed oxygen from the regulator.
   D. Bleed oxygen from the hoses and disinfect the entire tank and attachments.
19. What common error is made when placing a nonrebreathing mask on a patient?
   A. Bag is too small for the patient's face.
   B. Bag is full before placing the mask.
   C. Bag is not full before placing the mask.
   D. All of the above.

20. When should a nasal cannula be used?
A. When the patient is in full cardiac arrest
B. Only when the patient will not tolerate the nonrebreathing mask
C. Only on COPD patients and infants
D. Only on trauma patients

21. All patients who require oxygen should receive high concentrations of oxygen if they are cyanotic, cool, clammy, and short of breath.
A. True
B. False

22. Patients with COPD and infants who are cyanotic, cool, clammy, and short of breath should not receive high concentrations of oxygen.
A. True
B. False

23. Gastric distention (gastric inflation) is more common in which of the following?
A. Adults only
B. Children only
C. Infants only
D. Adults and children

24. Which of the following is not a special concern when an EMT uses a bag-mask device on a pediatric patient?
A. Avoid excessive bag pressure.
B. Use only enough to make the chest rise.
C. It may cause instant vomiting and aspiration.
D. Excessive bag-mask pressure, more than enough to make chest rise, can damage the lungs and inflate stomach.

25. Ordinary dentures should always be removed before ventilating a patient.
A. True
B. False

Section

3 Patient Assessment

Voices of Experience

Prepare For the Worst

It was a cold winter night and I was working on a transport ambulance. It was about 3:00 A.M. when the pagers went off. A van was on fire, possibly with someone inside. My ambulance arrived on the scene and parked at the curb downtown at the county courthouse. The local full-time fire department was already extinguishing the fire. I asked the fire fighter on the hose line if there was anyone inside the van. He said, "No." I then asked if anyone had checked, and he said that he was not sure.

At first, I assumed that since both departments were dispatched by the same dispatchers and should receive the same information, that the fire department had heard about the possible victim and a search had already been completed. It turned out that I was wrong. I spoke to about four fire fighters, and none of them could tell me if anyone had searched the vehicle.

Finally, on my request, one of the fire fighters searched the van and found a severely burned and unconscious male patient between the front bucket seats. He was removed quickly and we began treatment. He was severely and permanently damaged.

What had happened to cause such a lack of communication? The fire department was dispatched on a vehicle fire and nothing more. The ambulance was dispatched after the fire department when a second call came in that there was a possible victim inside the vehicle. The fire department and the ambulances were dispatched on different frequencies, so fire command did not hear the same information that we did. The miscommunication on this call resulted in an investigation, which led to changes in the dispatch system to help prevent a recurrence of this kind of problem.

For me, the lesson is never to assume that everyone is on the same page. If what is happening on the scene doesn't fit what you expect based on the information you have received, make the opposite assumption—that anything that can go wrong will go wrong. A true patient advocate hopes for the best, but expects the worst. Prepare for the worst and you can be more optimistic about your outcomes.

Brian F. Gainey
Fire Fighter/Paramedic
Training Officer
Greene County Ambulance Service
Bloomfield, Indiana

Patient Assessment

"The more we study the more we discover our own ignorance."
— Percy Bysshe Shelley

Section 3 covers the following subjects:

Lesson 3-1 Scene Size-Up

Enhance the EMT-B's ability to evaluate a scene for potential hazards, determine by the number of patients if additional help is necessary, and evaluate mechanism of injury or nature of illness. This lesson draws on the knowledge of Lesson 1–2.

Lesson 3-2 Initial Assessment

Provides the knowledge and skills to properly perform the initial assessment. In this session, the student will learn about forming a general impression, determining responsiveness, assessment of the airway, breathing, and circulation.

Lesson 3-3 Focused History and Physical Exam: Trauma Patients

Describes and demonstrates the method of assessing patients' traumatic injuries. A rapid approach to the trauma patient will be the focus of this lesson.

Lesson 3-4 Focused History and Physical Exam: Medical Patients

Describes and demonstrates the method of assessing patients with medical complaints or signs and symptoms. This lesson will also serve as an introduction to the care of the medical patient.

Lesson 3-5 Detailed Physical Exam

Teaches the knowledge and skills required to continue the assessment and treatment of the patient.

Lesson 3-6 Ongoing Assessment

Stresses the importance of trending, recording changes in the patient's condition, and reassessment of interventions to ensure appropriate care.

Lesson 3-7 Communications

Discusses the components of a communication system, radio communications, communication with medical direction, verbal communication, interpersonal communication, and quality improvement.

Lesson 3-8 Documentation

Assists the EMT-B in understanding the components of the written report, special considerations regarding patient refusal, the legal implications of the report, and special reporting situations.

Before assessing the patient, on arrival, an EMT has to assess what?

Scene safety.

What is a scene safety assessment?

An assessment to ensure the well-being of the EMT.

Describe four types of unsafe scenes.

Crash/rescue scenes; toxic substances—low oxygen area; crime scene potential for violence; unstable surfaces such as slope or those covered in ice or water.

In addition to protecting yourself, whom else do you have a duty to protect?

The patient and bystanders.

What is a scene size-up?

An assessment of the scene and surroundings that will provide valuable information to the EMT.

What do you do if your scene size-up reveals more patients than the responding unit can effectively handle?

Institute a mass casualty plan.

When should you obtain additional help?

Before contact with patients; you are less likely to call for help once you become directly involved in patient care.

After initiating a mass casualty plan and ensuring the scene is safe, what is your next responsibility?

To begin triaging patients.

In trauma, what important information can you gather from the scene?

Mechanism of injury.

What is the purpose of forming a general impression of the patient?

To determine priority of care; it is based on the EMT-B's immediate assessment of the environment and the patient's chief complaint.

What do you need to determine in your general impression of the patient?

Determine if ill (i.e., medical) or injured (trauma); if injured, identify mechanism of injury.

What additional information do you obtain in the general impression of the patient?

Age; sex; race.

What do you do if you determine that your patient has a life-threatening condition?

Treat immediately and assess the nature of the illness or mechanism of injury.

When first approaching a patient, what should you say?

Tell them your name, that you are an EMT, and that you are there to help.

What can you determine by asking the patient their name?

Airway and mental status.

What are the levels of mental status used by the EMT?

AVPU: Alert; responds to Verbal stimuli; responds to Pain; Unresponsive.

What is the quickest way to determine if the patient's airway is open?

The patient can talk or cry.

After ensuring an open airway, what do you need to assess next?

Is the patient breathing adequately?

You arrive at a playground and, as you approach the swing set, you see there is an approximately 4-year-old boy on the ground. His mother is shaking him and crying. What is the first thing you do?

Identify yourself; maintain spinal immobilization; assess airway.

The boy is unresponsive. What do you do?

Look, listen, and feel; if he is not breathing, use the jaw-thrust maneuver to open the airway while maintaining spinal immobilization.

After opening the child's airway, what do you do?

Look, listen, and feel for adequate breathing.

The child is breathing shallowly at 10 breaths per minute. What do you do?

Ventilate the patient via mouth-to-mask or bag-mask, with oxygen attached if available.

You begin ventilating the patient. What do you do now?

Check for adequacy of ventilation; check for pulse.

Your patient wakes up. He is breathing 50 breaths per minute. What should you do?

Place him on high-flow oxygen.

How is high-flow oxygen defined?

15 L/min via nonrebreathing mask.

When should you withhold oxygen to a patient with inadequate respirations?

Never.

Where do you palpate the pulse in a conscious adult?

Over the radial artery.

Where do you palpate the pulse in a child 1 year old or younger?

Over the brachial artery.

If there is no radial or brachial pulse, what do you do next?

Palpate the carotid pulse.

If there is no carotid pulse, what do you do next?

In a medical patient over 12 years old, start CPR and apply AED; in a medical patient less than 12 years old or a trauma patient, start CPR.

After starting CPR, what do you assess next?

Assess if major bleeding is present; if so, control bleeding. Assess perfusion.

How do you initially assess a patient's perfusion?

Skin color and temperature.

Where do you assess a patient's skin color?

Nail beds; lips; eyes.

What are the four abnormal skin color findings?

Pale; cyanotic; flushed; jaundiced.

What are abnormal skin findings on palpation?

Hot; cool; cold; clammy. Warm is normal.

What are you looking for in assessing the patient's skin condition?

The amount of moisture on the skin; abnormal is moist or wet.

In what kinds of patients is the capillary refill test most accurate?

Infants and children.

What is considered normal capillary refill time (CRT)?

Less than 2 seconds.

What kinds of patients fall under the category "priority"?

Poor general impression; unresponsive; responsive but not following commands; difficulty breathing; shock; complicated childbirth; chest pain with BP less than 100 systolic; uncontrolled bleeding; severe pain anywhere.

What should you consider when you have identified priority patients?

Expediting transport; ALS backup.

After completing your initial assessment, what should you do?

Proceed to the appropriate focused history and physical examination.

In the trauma patient, what is the main factor that will help you focus your history and physical exam?

Mechanism of injury.

The EMT curriculum lists 10 types of injuries it considers "significant." What are these?

Ejection; death in the same passenger compartment; falls of more than 20 feet; rollover; high-speed collision; vehicle–pedestrian collision; motorcycle crash; unresponsive or altered mental status; penetrations of the head, chest, or abdomen; hidden injuries (seatbelts and airbags).

Under what circumstances can airbags not be effective at preventing injury?

They may not be effective without a seatbelt; the patient can hit the wheel after deflation.

How can you check to see if the patient hit the steering wheel after airbag deflation?

Lift and look under the bag after the patient has been removed to see if there is damage to the steering wheel.

When should you perform the rapid trauma assessment on an otherwise well-appearing patient?

When there is a significant mechanism of injury.

As you inspect and palpate, what injuries or signs of injuries are you looking and feeling for?

Deformities; contusions; abrasions; punctures/penetrations; burns; lacerations; swelling; tenderness; instability; crepitus.

When assessing the neck of a trauma patient, what sign, in addition to the general trauma signs, are you looking out for?

Jugular vein distention (JVD).

When assessing the chest of a trauma patient, what sign, in addition to the general trauma signs, are you looking out for?

Paradoxical motion.

Where do you auscultate breath sounds in the trauma patient?

Apices; midclavicular line; at the bases; midaxillary lines.

What are you listening for when you auscultate breath sounds in the trauma patient?

Presence; absence; equality.

When examining the abdomen of a trauma patient, what specific qualities are you looking for?

Tenderness; firmness; softness; distention.

When examining the pelvis of a trauma patient, what specifically can you do to check for injury?

If no pain is noted, gently compress the pelvis to determine tenderness or motion.

When examining the extremities of the trauma patient, what are you checking for in addition to the general trauma signs?

Distal pulse; sensation; motor function.

Once you have assessed the anterior aspect of the trauma patient, what should you do?

Roll the patient using spinal precautions and assess the posterior body; inspect and palpate, examining for injuries or signs of injuries.

Following completion of the focused history and physical exam in trauma patients, what do you do next?

Obtain baseline vital signs and assess SAMPLE history.

For trauma patients with no significant mechanism of injury, such as a cut finger, what kind of history and physical exam do you perform?

Focused history and physical exam of injuries based on the components of the rapid assessment; obtain baseline vital signs and SAMPLE history.

What mnemonic is used in the EMT course to assess complaints and signs or symptoms, and what does it stand for?

OPQRST: Onset; Provocation; Quality; Radiation; Severity; Time.

After assessing complaints and signs and symptoms, what information do you elicit from the patient?

SAMPLE history.

In what direction do you do your rapid assessment?

From head to toe.

If the patient is unable to give you a SAMPLE history, how can you obtain it?
From bystanders, family, or friends.

When are you not required to complete a detailed physical exam?
When the patient has a specific injury, e.g., a cut finger.

When assessing the patient, what are you looking for?
Deformities; contusions; abrasions; punctures/penetrations; burns; tenderness; lacerations; swelling.

When examining the ears, what specifically are you looking for?
Drainage.

When examining the eyes, what specifically are you looking for?
Discoloration; unequal pupils; foreign bodies; blood in the anterior chamber.

When assessing the nose, what specifically are you looking for?
Drainage; bleeding.

When assessing the mouth, what specifically are you looking for?
Teeth; obstructions; swollen or lacerated tongue; odors; discoloration.

When assessing the neck, what specifically are you looking for?
Jugular vein distention; crepitus.

When assessing the chest, what specifically are you looking for?
Crepitus; paradoxical motion; breath sounds.

When assessing the abdomen, what specifically are you looking for?
Distention; firmness; softness.

When assessing the pelvis, what specifically are you looking for?
Pain on flexion and compression.

In assessing the extremities, what specifically are you looking for?
Distal pulses; sensation; motor function.

How often should you repeat your assessment in a stable patient?
Every 15 minutes.

How often should you repeat your assessment in an unstable patient?
Every 5 minutes.

What nine elements does the repeat assessment include?
Reassess mental status; maintain open airway; monitor breathing for rate and quality; reassess pulse for rate and quality; monitor skin color and temperature; reestablish patient priorities; repeat vital signs; repeat focused assessment regarding patient complaint or injuries; check interventions.

What is the term for a radio that is located at a stationary site, such as a hospital or public safety agency?
A base station.

At what wattage range do mobile two-way radios typically transmit?
20–50 watts

What is the typical transmission range for mobile two-way radios over average terrain?
10–15 miles.

What does a repeater do?
Receives a transmission from a low-power portable or mobile radio on one frequency and retransmits it at a higher power on another frequency.

What two recent additions have been made to the EMS communications systems, augmenting two-way analog radios?
Digital radio equipment; cellular telephones.

What agency is responsible for assigning EMS radio frequencies?
The Federal Communications Commission.

When responding to a dispatch, what do you need to inform the dispatcher on your callback?

That you received the call, are en route, and whether any other agencies need to be notified.

After acknowledging dispatch that you are en route, what is the next call you are required to make?

Arrival on scene.

True or false: Medical direction is located at the receiving facility.

False. In some systems it is at the receiving facility; in others, at a separate site.

What general characteristics are most important in your transmissions to medical control?

Need to be accurate, organized, concise, and pertinent.

What must you do after receiving an order or a denial from medical control?

Repeat the order back word for word.

When should you question orders received from medical control?

When they are unclear or appear to be inappropriate.

On the scene of a 55-year-old man complaining of chest pain, medical control orders you to give a tube of instant glucose sublingually. What should you do?

Question the order; glucose is not an appropriate medication.

In addition to receiving orders, what is the function of calling the receiving hospital?

It allows the staff to prepare for the arrival of your patient.

What is important to do before transmitting on the radio?

Make sure the volume is turned up; listen to ensure the frequency is clear.

How long should you wait after pressing the PTT button before speaking?

1 second.

How far away from the microphone should your lips be when speaking?

2–3 inches.

True or false: Always state your unit name and number at the beginning of the call up.

False. The name of the unit you are calling always comes first, then your unit name.

When you are busy and can't talk on the radio, how should you answer an incoming transmission.

Your unit name and the words "stand by."

True or false: When talking on the radio, you should vary your pitch and inflection.

False. Speak slowly, clearly, and in a monotone.

How often should you pause when delivering a lengthy transmission? Why?

Every 30 seconds; to allow emergency traffic to use the frequency if necessary.

True or false: Codes should always be used when available.

False. Avoid the use of codes.

Using phrases such as "be advised" adds to the professional tone of your radio report.

False. Avoid meaningless phrases that add nothing to your narrative.

True or false: It is essential to be courteous by using "please," "thank you," and "you're welcome" when using the radio.

False. Courtesy is assumed.

How can you be sure that numbers are correctly understood over the radio?

Give the number, then give the individual digits.

What is wrong with the phrase "Our ETA is approximately 15 minutes"?

ETA stands for estimated time of arrival; use of both "estimated" and "approximately" is redundant.

Why is it important not to give a patient's name over the air?

Airwaves are public; scanners are popular; it can violate patient confidentiality.

Why is it important to remain objective and impartial when describing patients over the air?
The EMT may be sued for slander if he or she injures someone's reputation on the air.

True or false: When describing the treatment you have completed, it is important to say "I" rather than "we."
False. An EMT-B rarely acts alone.

Why is it important to avoid profanity on the air?
It is unprofessional; the FCC may impose substantial fines.

What can you substitute for "yes" and "no"? Why should you do so?
"Affirmative" and "negative"; they are easier to hear and understand.

What should you say at the end of your transmission?
"Over" indicates the end of the transmission; wait for a confirmation to make sure the message was received.

True or false: Always end your report with a diagnosis of the patient's problem.
False. Most likely, you do not have sufficient information to make a diagnosis at this point; it is not part of EMT-B training or role to diagnose.

What can you do to reduce background noise when communicating en route?
Close the window; turn off other radios.

After notifying the dispatcher you are on scene, what is your next mandatory communication?
To notify the dispatcher you are leaving the scene.

What are the 11 elements of the verbal report to the receiving facility?
ID and level of provider; ETA; patient's age and sex; chief complaint; brief, pertinent history of the present illness; major past illnesses; mental status; baseline vital signs; pertinent findings of the physical exam; emergency medical care given; response to emergency medical care.

Under what circumstances must you call the receiving facility after your initial report?
If the patient's condition deteriorates or ETA changes.

What is the next mandatory call to dispatch after calling en route to the receiving facility?
Arrival at the receiving facility or in service.

After arrival at the receiving facility, what is the next mandatory call to dispatch?
Departing the receiving facility.

After notifying dispatch of your departure from the receiving facility, what is your next mandatory callout?
Arrival at the station.

What maintenance must be performed on communications equipment?
Periodic checks by a qualified technician.

What must an EMS system provide in the event of communications failure?
A back-up system for notifying dispatch and medical control.

What should you include in the verbal report to hospital staff on arrival?
Introduce the patient by name (if known); summarize the information given over the radio, including chief complaint; history that was not given previously; additional treatment given en route; additional vital signs taken en route; additional information that was collected but not transmitted.

True or false: It is important to make eye contact with your patients.
True.

True or false: You should never position yourself at a level lower than the patient.
False.

True or false: You should never frighten the patient with an honest assessment of the patient's condition.
False. Be honest with your patient.

True or false: Always use appropriate medical terminology when communicating with your patients.
False. Use language the patient can understand.

True or false: What you say is always more important than what is conveyed by your body language.

False. Always be aware of your own body language.

True or false: In what way can you speak to ensure that you are understood?

Speak clearly, slowly, and distinctly.

How should you address your patients?

Use proper name, either first or last, depending on the circumstances; ask the patient what he or she wishes to be called.

What can you do to help a patient who is having difficulty hearing you?

Speak clearly, with your lips visible.

What can you do to let the patient know what he or she says is important?

Allow the patient enough time to answer a question before asking the next one.

What can you do that will be most helpful in inspiring the confidence of your patient?

Act and speak in a calm, confident manner.

What must you keep in mind when communicating with the elderly?

The potential for a visual or auditory deficit.

What should you consider if it appears your patient is not understanding you?

A hearing impairment or a non-English speaker.

What seven elements are included in the "minimum data set" of patient information gathered at the time of the EMT's initial contact with the patient on arrival at scene, following all interventions, and on arrival at facility?

Chief complaint; level of consciousness (AVPU)—mental status; systolic blood pressure (patients over 3 years old); skin perfusion—capillary refill (for patients under 6 years old); skin color and temperature; pulse rate; respiratory rate and effort.

What times must be recorded in the "minimum data set" of administrative information?

Time incident reported; unit notified; arrival at patient; left scene; arrival at destination; transfer of care.

What are the functions of the prehospital care report?

Continuity of care; a legal document; educational; administrative; research; evaluation and continuous quality improvement.

What education function can the prehospital care report serve?

It can be used to demonstrate proper documentation and how to handle unusual or uncommon cases.

What administrative functions does the prehospital care report serve?

Billing and service statistics.

What two main types of prehospital run reports are there?

Traditional written form with check boxes and a section for narrative; computerized versions.

What five sections are part of the typical prehospital run report?

Run data; patient data; check boxes; narrative section; other state or local requirements.

What information is included in the patient data section of the prehospital run report?

Patient name; address; DOB; insurance information; sex; age; nature of call; mechanism of injury; location of patient; treatment prior to arrival; signs and symptoms; care administered; baseline vital signs; SAMPLE history; changes in condition.

True or false: The narrative portion of the prehospital run report should provide your conclusions and diagnosis.

False. Describe, don't conclude.

True or false: The narrative portion of the prehospital run report should include pertinent negatives.

True.

True or false: You should use radio codes in the narrative section of the prehospital run report.

False.

When should you use abbreviations in the narrative section of the prehospital run report?
If they are standard.

When is it vital to document the source of information on the prehospital run report?
When that information is of a sensitive nature, e.g., communicable diseases, information with legal ramifications.

True or false: The prehospital run report is covered under sunshine laws and must be made available when requested.
False. It is confidential, be familiar with state laws.

How can falsification of a prehospital run report lead to poor patient care?
Because other health care providers have a false impression of which assessment findings were discovered or what treatment was given.

How should an EMT document an error of omission or commission?
Document what did or did not happen and what steps, if any, were taken to correct the situation.

Who may refuse treatment?
A competent adult patient.

True or false: Before leaving the scene, an EMT should try to persuade a patient who is refusing transport to go to the hospital.
True.

What factors may make an adult incompetent to refuse treatment?
Alcohol; drugs; illness; or injury may impair judgment.

True or false: Medical control should be consulted when a patient refuses treatment.
True.

True or false: Once you have begun treating a patient, he or she may not refuse transport.
False.

What should you document when a patient refuses transport?
Any assessment findings and emergency medical care given; have the patient sign a refusal form.

Who should witness a refusal of treatment form?
A family member, police officer, or bystander.

What should you do if the patient refuses to sign a refusal of treatment form?
Have a family member, police officer, or bystander sign the form verifying that the patient refused to sign.

True or false: A refusal of transport form stands in lieu of a prehospital run report.
False.

What should be documented on the prehospital run report when a patient refuses treatment?
A complete patient assessment; care provided; statement that the EMT explained to the patient the possible consequences of failure to accept care, including potential death; offer of alternative methods of gaining care; that the EMT stated a willingness to return if the patient calls again for further assistance.

How do you correct an error on the prehospital run report while the form is being written?
Draw a single horizontal line through the error, initial it, and write the correct information beside it.

Why should you not try to erase an error you have just made on a run report?
It may be interpreted as an attempt to cover up a mistake.

How should you correct an error after the report form has been submitted?
Preferably in different color ink, draw a single line through the error, initial and date it, and add a note with the correct information.

How should you correct an omission in a report already submitted?
Add a note with the correct information, the date, and your initials.

Should the standard for completing prehospital run reports in a mass casualty incident (MCI) be the same as for a typical call?
No. The local MCI plan should have special guidelines.

What may help you in organizing information for report writing after a mass casualty incident?

Triage tags and a special system for recording information temporarily during MCIs.

When should special situation reports be completed?

When you have to document events that should be reported to local authorities or to amplify and supplement the primary report.

How long after an incident should a special situation report be filed?

In a timely manner; as soon as possible.

What types of situations call for a special situation report?

Exposures; injuries; criminal activity.

How can the prehospital run report serve a continuous quality improvement program?

Information gathered from run reports can be used to analyze various aspects of the EMS system; they can then be used to improve different components of the system and prevent problems from occurring.

Patient Assessment

1. On arrival, before assessing the patient, an EMT must assess what?
   A. Vital signs
   B. Hazardous material spills
   C. Scene safety
   D. Breath sounds, rate, and quality

2. In addition to protecting yourself, who else do you have a duty to protect?
   A. Only the patient
   B. Your partner
   C. All EMTs and rescuers on scene
   D. The patient and bystanders

3. After initiating a mass casualty plan and ensuing that the scene is safe, what is your next responsibility?
   A. Order all resources.
   B. Begin triaging patients.
   C. Establish a command post.
   D. Begin CPR on priority-one patients.

4. Scene safety assessment is a/an:
   A. Good-faith effort to protect the patient.
   B. Good-faith effort to protect the patient and bystanders.
   C. Assessment to ensure the well-being of the EMT.
   D. Assessment to know what is safe to do for the patient.

5. What can you determine by asking the patient his or her name?
   A. Mental status only
   B. Airway and mental status
   C. Responsiveness of victim
   D. If the patient is breathing adequately

6. What should the EMT do after opening the airway of a child?
   A. Call for help.
   B. Look, listen, and feel for adequate breathing.
   C. Look and feel for a pulse.
   D. Turn on and use a nasal cannula.

7. For a conscious adult, you should palpate the pulse:
   A. Over the radial pulse.
   B. Always using the carotid pulse only.
   C. Over the brachial pulse/artery.
   D. Over the pedal pulse.

8. What do you do next for an adult or child if there is no radial or brachial pulse?
   A. Begin rescue breaths.
   B. Begin CPR.
   C. Palpate the carotid pulse.
   D. Palpate the pedal pulse.

9. You assess a patient's skin color by looking at the:
   A. Nail beds.
   B. Gums and lips.
   C. Nail beds, lips, and eyelids.
   D. Palms of hands and the inside of the nasal opening.

10. When an EMT checks a capillary refill test (CRT), what is considered normal capillary refill?
A. 2 seconds or less
B. 5 seconds or less
C. 10 seconds or less
D. 15 seconds or less

11. The EMT curriculum lists general types of injuries that are considered to be "significant." Ejection from a vehicle, rollover vehicle collisions, falls of 20 feet or three times the victim's height, and altered mental status are all considered to be significant.
A. True
B. False

12. After assessing complaints and signs and symptoms, what information do you elicit from the patient?
A. DOTS
B. SAMPLE history
C. Medical history
D. OPQRST

13. How can the SAMPLE history be obtained when the EMT is unable to obtain it from the patient?
A. From the mechanism of injury
B. From the person who called 9-1-1
C. From bystanders, friends, or family
D. By assessing the AVPU scale

14. When an EMT assesses the patient's ears, he or she is specifically looking for drainage. When assessing the patient's nose, what specifically is the EMT looking for?
A. Drainage only
B. Drainage and bleeding
C. Crepitus and drainage
D. Foreign bodies and bleeding

15. What should you do if medical control orders a treatment that you know is not appropriate for the patient?
A. You must follow the order.
B. Tell medical control that you question the order.
C. Have your partner call the hospital and speak to the charge nurse via cell phone and ask her to reverse the order.
D. Follow the order; then document that you question what you were told to do.

16. It is essential to be courteous by using "please," "thank you," and "you are welcome" when using the radio.
A. True—courtesy is essential.
B. False—courtesy is assumed.

17. During a radio transmission, the EMT should substitute for "yes" and "no." What substitutes should you use?
A. "Yes, sir" or "yes, ma'am"
B. "No, sir" or "no, ma'am"
C. "Affirmative" or "negative"
D. "Negative" or "assuredly"

18. It is important not to give a patient's name over the air because the airwaves are public and:
A. Scanners are popular.
B. It can violate patient confidentiality.
C. Both A and B are correct.
D. Only B is correct.

19. You should never frighten the patient by acting unprofessionally. For example, do not say something like, "Oh my goodness, your whole leg is torn right off." However, you must:
A. Be honest with your patient.
B. Never tell a patient how serious an injury is.
C. Not tell the patient, but warn a family member or friend what is wrong.
D. Both A and B are correct.

20. When might you encounter a patient with the potential for a visual or auditory deficit?
A. Preteen
B. Young, frightened teenager
C. Any adult
D. Elderly patient

21. What can the EMT do that will be most helpful in inspiring the confidence of the patient?
A. Show humility.
B. Show compassion.
C. Act and speak in a calm, competent manner.
D. Act quickly and display that you know what you are doing.

22. The time that the incident was reported, unit notified, arrival at patient, left scene, arrival at destination, and transfer of care are important pieces of administrative information that must be recorded:
A. By the EMT before, during, and after treatment.
B. In the minimum data set.
C. In the maximum data set.
D. By the dispatcher on the prehospital care report.

23. How should an EMT document an error of omission or commission?
A. Document only errors of commission.
B. Document only errors of omission.
C. Document errors of omission or commission only when told to do so by a supervisor.
D. Document what did or did not happen and what steps, if any, were taken to correct the situation.

24. Only a competent adult patient may refuse treatment.
A. True
B. False

25. Exposures, injuries, and criminal activity often require the EMT to:
A. Stop administering care and deal with that situation immediately.
B. Prepare a special-situation report.
C. Document the information on the prehospital care report.
D. None of the above

Section

Medical Emergencies

Voices of Experience

Wrestling in the Dark

I guess you could say it was a dark and stormy night and be right. I was a new fire fighter at a small, volunteer fire department that covered an area of about 60 square miles and served about 12,000 residents. I was also a brand-new EMT-B, one of only a handful of EMTs in northwestern Indiana in 1996. Our department provided BLS nontransport first response only. We needed an ALS ambulance from another service to transport our patients. These ALS units were anywhere from 15 to 40 minutes away, even though they were dispatched at the same time we were.

During my time at this department, it was not uncommon to leave one scene and immediately be dispatched to another. Such was the case on this night.

Responding from the scene of another call, I was called to the scene of a diabetic man with altered mental status. Each EMT was issued an EMS bag, and we responded in our personal vehicles with a blue light on the roof.

While en route to the scene, I heard our department dispatched on a fire, and all the other fire department personnel responded to it. I now knew that I would be alone on this run, with no backup, with a 30-minute wait for an ALS unit.

I pulled up at the address, a small farmhouse not more than a mile from our firehouse. I was greeted by an elderly lady who said that her husband was a diabetic and wasn't getting up from his bed for dinner. As I walked in, I heard the fire trucks pass by the residence on their way to the other run. I looked around the house and saw no apparent danger. I assumed this was to be a routine run. Yeah, I know what you're thinking—never assume.

I walked down the narrow hallway to the patient's bedroom. The door was partially closed, but I could see light from around the doorway. I identified myself prior to entering. "Fire department. My name is Rob and I am an EMT. Your wife called and said that you needed some help." I opened the door and, to my surprise, I found an elderly male lying naked, supine on the bed, covered in sweat.

He immediately looked at me, and I noticed that his eyes seemed to stare off into space, as if the lights were on but nobody was home. He was reaching for a basket next to the nightstand. In the basket, I saw a shotgun, a broken baseball bat, and a rifle. His wife, who followed me into the room, screamed, and ran out of the room. I was now stuck in a small bedroom with an altered mental status patient reaching for a basket of weapons. Houston, we have a problem.

I had no time to turn around and flee the room, because he already had his hand on the shotgun and was pulling it out of the basket. I dropped my EMS bag, dove onto the bed, and was now wrestling with him for control of the gun. I could not reach my radio to call for backup or police assistance. In the midst of this life-and-death struggle, I tried to tell the patient that I was an EMT and was not there to harm him. I should add that I am no ninety-eight-pound weakling. This patient may have been getting on in years, but he was a 200-plus-pound farmer, and those pounds were mostly muscle.

We struggled for the next 20 or 30 minutes. It seemed like forever. I finally got the shotgun away from him and he immediately reached for the baseball bat. During this whole battle, his wife did not help or call for the police, despite my calls for her to do so. As we wrestled, he could not speak intelligibly, but instead had garbled responses to my pleas for him to let go of the gun. He grabbed the broken bat, but I was able to pull it from his hands. I was now trying to keep him from hitting me with his fists while I tried to make my escape. As I threw the gun down the hallway and used the bat to push him away from me, it was with great relief that I heard the siren of the ALS unit approaching. When the medics entered, I yelled for help. I yelled that I had gotten a gun away from the patient and that we needed the cops right now!

The medics called for the police and helped me restrain the patient. They checked his blood sugar after I told them he was combative and a diabetic. They then gave him D_{50} through an IV that they got in his arm while I was lying on top of him to pin him between the wall and the bed. As his blood sugar began to rise, he stopped fighting. He was confused as to why all these people were in his house and why his gun and bat were on the floor. Finally, I left the room and found my radio in the hallway where it had landed sometime during the melee. I picked up my EMS bag and walked out to the ambulance in the driveway. That's when it finally dawned on me how close to death I had come.

I was shaking as I helped the medics load the patient into the ambulance. As they left for the hospital, the patient's wife said, "Thanks for your help. I told you he wasn't acting right." I just shook my head, got back into my car, and headed for the station to write my report.

When I got to the station, I learned that the fire run that went out after my call was a false alarm, and that the other fire fighters had been in service for about 25 minutes. As I walked in, the guys at the station jokingly asked if I had needed any help on my run. I told them about my ordeal, and they were shocked and felt really bad that they hadn't helped.

As you can imagine, this call changed the way I approach scene safety and the combative diabetic patient. No matter how cautious you think you may be, scenes can change rapidly and without warning. Diabetic patients with low blood sugar can be very combative and very strong. No matter how much you try to explain things, no matter how you try to calm them, they cannot be rational until you raise their blood sugar to a level at which their brains can function. Enter every house, every apartment, and every room as if somebody dangerous is inside. Always leave yourself a way out—and be ready to use it. Never let a door close behind you and never let somebody come between you and your path to safety.

Robert C. Allen, BS, AAS, NREMT, PI
Lieutenant, Benton Township Volunteer Fire Department
Unionville, Indiana
Fire Fighter/Paramedic, Sugar Creek Township Fire Department
New Palestine, Indiana

SECTION

4 Medical Emergencies

"I am dying from the treatment of too many physicians."
— Alexander the Great

Section 4 covers the following subjects:

Lesson 4-1 General Pharmacology

Provides the student with a basic knowledge of pharmacology, providing a foundation for the administration of medications given by the EMT and those used to assist a patient with self-administration.

Lesson 4-2 Respiratory Emergencies

Reviews components of the lesson on respiratory anatomy and physiology. It will also provide instruction on assessment of respiratory difficulty and emergency medical care of respiratory problems and the administration of prescribed inhalers.

Lesson 4-3 Cardiovascular Emergencies

Reviews the cardiovascular system, an introduction to the signs and symptoms of cardiovascular disease, administration of a patient's prescribed nitroglycerin, and use of the automated external defibrillator.

Lesson 4-4 Diabetes/Altered Mental Status

Reviews the signs and symptoms of altered level of consciousness, the emergency medical care of a patient with signs and symptoms of altered mental status and a history of diabetes, and the administration of oral glucose.

Lesson 4-5 Allergies

Teaches the student to recognize the signs and symptoms of an allergic reaction, and to assist the patient with a prescribed epinephrine autoinjector.

Lesson 4-6 Poisoning/Overdose

Teaches the student to recognize the signs and symptoms of poisoning and overdose. Information on the administration of activated charcoal is also included in this section.

Lesson 4-7 Environmental Emergencies

Covers recognizing the signs and symptoms of heat and cold exposure, as well as the emergency medical care of these conditions. Information on aquatic emergencies and bites and stings will also be included in this lesson.

Lesson 4-8 Behavioral Emergencies

Develops the student's awareness of behavioral emergencies and the management of the disturbed patient. Restraining the combative patient will also be taught in this lesson.

Lesson 4-9 Obstetrics/Gynecology

Reviews the anatomical and physiological changes that occur during pregnancy, demonstrates normal and abnormal deliveries, summarizes signs and symptoms of common gynecological emergencies, and neonatal resuscitation.

There will be specific circumstances in which an EMT will need to assist a patient in administering his or her own medication. It will be important for the EMT to have a basic understanding of the proper names of medications. With that in mind, where can you research a medication's generic name?

In the *U.S. Pharmacopoeia*, a governmental publication listing all drugs in the United States.

Each medication has a specific set of indications. What are "indications"?

The most common uses for the drug in treating a specific illness.

Each medication has a specific set of contraindications. What are "contraindications"?

Situations in which a drug should not be used because it may cause harm to the patient or offer no effect in improving the patient's condition or illness.

Explain the meaning of the word "dose."

The amount of the drug that should be administered.

Prior to an EMT giving a patient any medication, he or she should understand that drug's action. Explain what is meant by a drug's "action."

The desired effect a drug has on the patient and/or body system.

Prior to an EMT giving a patient any medication, he or she should understand that drug's side effects. Explain what is meant by a drug's "side effects."

Any action of a drug other than the desired action(s); some side effects may be predictable.

For years, EMTs have been administering a medication in a variety of situations that has helped thousands of patients. Identify that medication.

Oxygen.

List the major structures of the respiratory system.

Nose; nasopharynx; mouth; oropharynx; larynx; trachea; bronchi; lungs.

List the signs of adequate air exchange.

The patient has breath sounds that are clear bilaterally; skin is pink, warm, and dry; no use of accessory muscles; no sternal retractions; the chest wall expands evenly; breathing rate is within normal limits; the patient appears calm; the patient is able to speak in complete sentences without gasping for air.

These structures are the two major branches of the trachea to the lungs. What are they?

Bronchi.

What is the normal breathing rate for a healthy adult at rest?

12–20 breaths/min.

What is the normal breathing rate for a healthy child at rest?

15–30 breaths/min.

What is the normal breathing rate for a healthy infant at rest?

25–50 breaths/min.

Veronica, an experienced EMT, is treating a child who is experiencing moderate dyspnea. What are the concerns in dealing with this type of patient?

A child with respiratory distress will deteriorate into respiratory failure and circulatory collapse, eventually resulting in respiratory arrest if left untreated.

Explain why oxygen is beneficial to children in respiratory distress.

Use of oxygen can block the progression of respiratory distress leading to respiratory arrest and may even reverse it to some degree; when possible, deliver humidified oxygen and allow the child to remain in the caregiver's lap.

Identify all the contraindications of giving high-concentration oxygen to infants and children.

NO contraindication to giving high-concentration oxygen to infants and children.

Elena and her partner, Sam, are treating a victim of a gunshot wound. What are some indications that Sam is adequately artificially ventilating the patient?

The chest rises and falls with each artificial ventilation; the rate is sufficient, approximately 12 times per minute; the heart rate returns to normal with successful artificial ventilation.

Mike is in the back of the rig assessing a patient complaining of a breathing problem. Identify some signs and symptoms of breathing difficulty.

Shortness of breath, restlessness, increased pulse rate, increased breathing rate, decreased breathing rate.

The EMT must be able to identify a patient who is experiencing breathing difficulty. With that ability, the EMT must be able to identify noisy breathing. Give several examples of "noisy" breathing.

Can be classified by any and all of the following: crowing, audible wheezing, gurgling, snoring, and stridor.

What is stridor?

A harsh sound heard during breathing that indicates the patient has an upper airway obstruction.

A patient with prolonged breathing difficulty may begin having retractions to aid her respiratory effort. What is meant by "retractions"?

A term used when a patient begins using accessory muscles to aid his or her respiratory effort.

Patients may sit in a tripod position to aid in their respiratory effort. Describe the "tripod" position.

When the patient positions him- or herself in a sitting position with feet dangling and leaning forward; this position assists the dyspneic patient by easing respiratory effort.

Captain Haskell is assessing a patient complaining of difficulty breathing. He is attempting to perform a focused history and physical exam. Identify important questions he should ask the patient in order to obtain a complete history.

When was the onset of this current set of symptoms? What in particular was the provocation of the current incident? What is the quality of the patient's respirations? Is there any radiation of pain or discomfort? How would the patient compare the severity of the current condition to prior episodes? How long has the patient been experiencing difficulty? Has the patient attempted any interventions to help him- or herself feel better?

What treatment should the EMT use initially if a patient complains of dyspnea?

Apply oxygen to the patient with the proper device; obtain a baseline set of vital signs.

Patients often have medications at home available to them in the event of respiratory distress. They are packaged in both the generic and trade names. List several prescribed inhalers that a patient may have by the generic names.

Generic names could be any or all of the following: albuterol, isoetharine, and metaproteranol.

Patients often have medications at home available to them in the event of respiratory distress. They are packaged in both the generic and trade names. List several prescribed inhalers that a patient may have by the trade names.

Trade names could be any or all of the following: Proventil, Ventolin, Bronkosol, Bronkometer, Alupent, and Metaprel.

List the indications for using a handheld inhaler.

The use of a handheld inhaler by an EMT must meet all of the following criteria: the patient must exhibit signs and symptoms of a respiratory emergency; the patient must have a physician-prescribed handheld inhaler; the EMT must have specific authorization from medical control.

List the contraindications for using a handheld inhaler.

All of the following are contraindications for using a handheld inhaler by the EMT in the prehospital setting: the patient is unable to use the inhaler as a result of his or her symptoms; the inhaler is not prescribed for the patient; the EMT does not have permission from medical control to use the device; the patient has already met the maximum prescribed dose prior to the arrival of the EMT.

What is meant by the sentence "Make sure that you give the patient the proper dosage"?

The dosage a patient receives from a handheld inhaler is the number of inhalations based on medical direction's order or physician's order based on consultation with the patient.

Describe the steps necessary for the EMT to assist a patient with a handheld inhaler.

The EMT should obtain the order for administration from medical control; assure the right medication, right patient, right route, and that the patient is alert enough to follow instructions; check the expiration date of the inhaler; check to see if the patient has already taken any doses; assure the inhaler is at room temperature or warmer; shake the canister vigorously several times; remove the patient from the oxygen supply briefly; instruct the patient to exhale deeply and place his or her lips around the opening of the inhaler; have the patient depress the handheld inhaler as he or she begins to inhale deeply; instruct the patient to hold his or her breath for as long as he or she comfortably can so that the medication can be absorbed; replace oxygen on the patient; allow the patient to breathe a few times and repeat a second dose per medical direction orders.

A patient may have a "spacer device" for use with their inhaler. What is it and what is its function?

An attachment between the inhaler and the patient; it allows for more effective use of the medication.

What physiological actions are we as EMTs looking for when we use handheld inhalers of Beta agonist bronchodilators?

To dilate bronchioles reducing airway resistance.

What possible side effects can occur with the use of handheld inhalers?

Any or all of the following side effects may be seen in a patient using an inhaler: pulse rate may increase; the patient may experience tremors; nervousness.

Describe the reassessment strategies that the EMT may use in the field to monitor the patient's progress.

Gather an additional set of vital signs and repeat the focused assessment.

What special considerations should the EMT think about while evaluating infants and children?

The EMT should remember that the use of handheld inhalers is very common in children; retractions are more commonly seen in children than in adults and cyanosis is a late finding in children; very frequent coughing may be present rather than wheezing in some children; the emergency medical care with usage of handheld inhalers is the same if the ill child meets the indications for usage of the inhalers.

At the morning shift change, Lieutenant Blazer tells you that you are responsible to teach today's EMS drill covering diabetic emergencies. He instructs you to prepare a lesson for the rest of the crew covering the signs and symptoms associated with a patient with altered mental status who has a history of diabetes controlled by medication. Describe the information you are going to present to the rest of the crew.

Rapid onset of altered mental status because of lowered blood sugar may occur in the following situations: the patient missed a meal on a day he or she took the prescribed insulin; the patient vomited a meal after taking insulin; the patient may experience altered mental status after an unusual exercise or physical work episode.

Describe how a patient suffering from hypoglycemia may physically appear to the responding EMT.

The patient may have any or all of the following: an intoxicated appearance; staggering; slurred speech; complete unresponsiveness; an elevated heart rate; cold, clammy skin; hunger; seizures.

Many patients take an oral medication to assist in controlling their diabetes. Name three of those medications.

Diabinese; Orinase; Micronase.

What kinds of behavior can be exhibited by a patient suffering from hypoglycemia that can be confused with psychological disturbances?

Anxiety or combativeness.

Should all seizures in children be considered life threatening?

Seizures in children who have chronic seizures are rarely life threatening; however, undiagnosed seizures, including febrile, should be considered life threatening by the EMT.

Identify potential causes of seizures.

Seizures may be caused by any or all of the following: fever; infections; poisonings; hypoglycemia; trauma; hypoxia; they may be idiopathic in children.

Describe the emergency medical treatment for a patient having a seizure.

Assure the airway is patent; position the patient so he or she will not injure him- or herself, preferably on his or her side if there is no chance of any spinal injury; have suction ready; provide supplemental high-flow oxygen; transport as soon as possible to a hospital.

List potential conditions that could cause a patient to have an altered mental status.

Hypoglycemia; poisoning; postseizure; infections; head trauma; hypoxia.

Describe the basic emergency medical care for a patient with an altered mental status not related to hypoglycemia.

Ensure a patent airway; provide supplemental high-flow oxygen; be prepared to suction; assist with ventilation if necessary; transport to the hospital as soon as possible.

Describe the basic emergency medical care for a patient with an altered mental status related to hypoglycemia.

Perform an initial assessment with a focused history and physical exam; determine onset and duration of the incident and any associated symptoms; look for any evidence of trauma, seizures, or fever; obtain baseline vital signs; determine last meal, last medication dose, and any related illness; determine if patient can swallow; administer oral glucose in accordance with local or state medical direction or protocol.

Identify two trade names for oral glucose.

Glutose; Insta-Glucose.

What contraindications does the EMT need to be concerned with in regard to the administration of oral glucose?

Patient is unresponsive and unable to swallow.

Describe the proper procedure used to administer a tube of oral glucose to a patient presenting with an altered mental status related to hypoglycemia.

Ensure signs and symptoms of altered mental status with a known history of diabetes; obtain an order from medical control; assure patient is conscious and can swallow and protect the airway; administer glucose between cheek and gum; perform an ongoing assessment, looking for change in mental status.

What "action" does oral glucose have on a patient?

Increases a patient's blood sugar for proper function of the brain.

What are the side effects of administering oral glucose?

No negative side effects of giving a patient oral glucose when administered properly. Caution should be paid to ensure that a patient does not aspirate the medication; this can occur in a patient without a gag reflex.

What is the definition of an allergic reaction?

An exaggerated immune response to any substance.

What are some possible causes of an allergic reaction?

Insect bites or stings; foods, especially nuts or seafood; plants; medications.

A patient suffering from an allergic reaction may have any or all of the following signs and symptoms. List the potential signs and symptoms that may be found when assessing the patient's skin.

The patient may have a warm tingling feeling in the face, mouth, chest, feet, and hands; he or she may have itching, hives, red skin (flushing), and swelling to the face, neck, hands, feet, and tongue.

A patient suffering from an allergic reaction may have any or all of the following signs and symptoms. List the potential signs and symptoms that may be found when assessing the patient's respiratory system.

The patient may feel tightness in the throat or chest; may cough; may have rapid and/or labored breathing; respirations may be noisy with hoarseness, stridor, or wheezing.

A patient suffering from an allergic reaction may have any or all of the following signs and symptoms. List the potential signs and symptoms that may be found when assessing the patient's cardiac system.

The patient's heart rate may be increased; his or her blood pressure decreased.

A patient suffering from an allergic reaction may have any or all of the following signs and symptoms. List the potential signs and symptoms that may be found when assessing the patient's overall general condition.

Itchy, watery eyes; headache; runny nose; sense of impending doom; decreasing mental status.

What additional findings could lead you to believe a patient may be experiencing an allergic reaction?

Assessment findings that reveal shock or respiratory distress indicate the presence of a severe allergic reaction.

Describe the emergency medical care for a patient who has come in contact with a substance that has caused an allergic reaction and who complains of respiratory distress or exhibits signs and symptoms of shock.

Perform an initial assessment to include ABCs; perform a focused history and physical exam; determine any history of allergies and what the patient was exposed to; determine how the patient was exposed (inhalation, sting, oral, etc.); assess baseline vital signs; administer high-flow oxygen with a nonrebreathing mask; determine if the patient has a prescribed preloaded epinephrine pen available; facilitate administration of the preloaded epi-pen; contact medical control and reassess in 2 minutes; if the patient does not have an epinephrine auto-injector available, transport immediately.

Describe the emergency medical care of a patient who has had contact with a substance that has caused an allergic reaction without signs of respiratory distress or shock.

Perform an initial assessment to include ABCs; perform a focused history and physical exam; determine any history of allergies and what the patient was exposed to; determine how the patient was exposed (inhalation, sting, oral, etc); assess baseline vital signs; administer low-flow oxygen; if the patient is not wheezing or is without signs of respiratory compromise or hypotension, he or she should not receive epinephrine.

Lieutenant Pete Jaegly of the TFD is assessing a patient who stated she took a new medication and is having an allergic reaction. Why is it important for the lieutenant to closely monitor this patient's airway?

The patient may initially present with airway/respiratory compromise, or this compromise may develop as the allergic reaction progresses.

There are three criteria that must be met prior to using an epinephrine auto-injector. What are they?

The patient must be exhibiting the signs and symptoms of an allergic reaction; the medication is prescribed for this patient by a physician; medical direction authorizes its use for this patient.

Are there any contraindications to using an epinephrine auto-injector?

Not when used in a life-threatening situation.

What is the dosage for an adult and child epinephrine auto-injector?

The proper dose for an adult is one adult auto-injector containing 0.3 mg of epinephrine; the proper dose for a child/infant is one child/infant auto-injector containing 0.15 mg of epinephrine.

Describe the proper procedure for using an epinephrine auto-injector.

Remove the safety cap and place the auto-injector against the lateral portion of the thigh, midway between the waist and the knee; push the injector firmly against the thigh until the injector activates; hold the injector in place until the medication is injected and record the time; dispose of the auto-injector in a biohazard container.

What "action" does epinephrine have on a patient?

Epinephrine's actions include both dilating the bronchioles to aid respiration and constricting the blood vessels to maintain blood pressure.

Several side effects could be associated with using an epinephrine auto-injector. What are they?

Any or all of the following side effects could be present: increased heart rate; pallor; dizziness; chest pain; headache; nausea and vomiting; excitability and anxiousness.

The potential exists that, despite using an epinephrine auto-injector for a patient with an allergic reaction, his or her condition continues to worsen. What should you be continually assessing to monitor this patient's condition?

Increasing breathing difficulty; decreasing mental status and blood pressure.

If a patient does not improve after using an epinephrine auto-injector, what should be your next action?

Obtain medical direction; give an additional dose of epinephrine with an auto-injector; treat for shock; administer oxygen; prepare to initiate basic cardiac life-support measures, which should include CPR and the use of an AED if indicated.

If a patient does improve after using an epinephrine auto-injector, what should be your next action?

After completing a reassessment, provide oxygen, supportive care, and transport to the hospital.

In assessing a patient who has been potentially poisoned, what questions should be asked of the patient?

What is the substance you may have been poisoned with? When did you become exposed to or ingest this poison? How much did you ingest? Over what length of time? How much do you weigh?

What are the signs and symptoms of an ingested poison?

Nausea; vomiting; diarrhea; altered mental status; possible abdominal pain; chemical burns around the mouth.

What is the proper treatment of the ingestion of a poison?

Complete assessment of the ABCs; a brief history and physical exam; remove pills, tablets, or fragments with gloves from the patient's mouth, as needed, without personal injury; consult medical control and consider the use of activated charcoal; bring all containers, bottles, labels, etc., of poison agents to the receiving facility.

What are the signs and symptoms of inhaled poisons?

Difficulty in breathing; chest pain; cough; hoarseness; dizziness; headache; confusion; altered mental status; seizures.

What is the proper emergency medical care for a patient who has inhaled a poisonous substance?

Have trained rescuers remove the patient from the poisonous environment; give high-flow oxygen with a nonrebreathing mask; complete an assessment of the ABCs and obtain vital signs; bring all containers, bottles, labels, etc., of poison agents to the receiving facility if it can be done safely.

What are signs and symptoms of injection with a toxic substance?

Weakness; dizziness; chills; fever; nausea; vomiting.

What basic emergency medical care should Lieutenant Mattox provide for the patient who has been injected with a poisonous substance?

Provide support for the patient's airway and provide oxygen; be alert for vomiting; bring all containers, bottles, labels, etc., of poison agents to the receiving facility if it can be done safely.

Poisons can also be absorbed into the body. List the signs and symptoms of a patient who may have absorbed a poison into his or her body.

There may be a liquid or powder on the patient's skin; skin may be burned or irritated with redness and itching.

What is the proper emergency care for a patient who has absorbed a poison into his or her body?

Remove contaminated clothing while protecting yourself from exposure to the poison; brush off any visible powder; if a liquid is encountered, irrigate with clean water for at least 20 minutes; if the poison involves the eyes, irrigate with clean water for at least 20 minutes and continue en route to the hospital.

Activated charcoal is used in the treatment of certain poison victims. What are the indications for use by the EMT?

Activated charcoal is indicated for the ingestion of certain poisons by mouth.

Activated charcoal is used in the treatment of certain poison victims. What are the contraindications for use by the EMT?

Contraindications for the use of activated charcoal include a patient who may have an altered mental status; have ingested acids or alkalis; and is unable to swallow.

How is activated charcoal packaged for use by an EMT?

Commonly premixed in water, frequently in plastic bottles containing 12.5 grams of activated charcoal; powder should be avoided in the field, as it is messy and difficult to mix.

What is the correct dose of activated charcoal that should be given to an adult?

25–50 grams.

What is the correct dose of activated charcoal that should be given to a child?

12.5–25 grams.

Describe the proper procedure for administering activated charcoal to a patient.

Obtain an order from medical control; the container must be shaken thoroughly; because the medication looks like mud, the patient may need to be persuaded to drink it; record the time it is completely in the patient's stomach.

What is the purpose of giving activated charcoal to a patient who has ingested a poison?

Activated charcoal binds to certain poisons and prevents them from being absorbed into the body.

What are the potential side effects of giving a patient activated charcoal?

The patient may have black stools; some patients who have ingested poisons that cause nausea may vomit. If that occurs, you may repeat the dosage once.

An EMT should be able to distinguish between behavior that is acceptable and that which is not. Describe what type of behavior would be considered a behavioral emergency.

A situation in which the patient exhibits abnormal behavior within a given context that is unacceptable or intolerable to the patient, family, or community; behavior can be due to extremes of emotion leading to violence or other inappropriate behavior or due to a psychological or physical condition such as lack of oxygen or low blood sugar in diabetics.

A number of general factors may influence a person's behavior. Identify some common causes of behavior alteration.

Situational stresses; medical illness; psychiatric problems; alcohol or drugs.

Several other specific influences can affect a person's behavior. Identify some of these causes of behavior alteration.

The patient may be hypoglycemic, hypoxic, have diminished blood flow to the brain, or have head trauma; the patient may also be experiencing a psychogenic condition resulting in psychotic thinking, depression, or panic. Consider excessive cold or hot temperatures.

A patient may be experiencing a psychological crisis resulting in behavior alteration. What are the symptoms an EMT should be looking for?

Panic, agitation, bizarre thinking and behavior; the patient may be a danger to him- or herself or others; could result in suicide or violence to others.

When confronted with a patient who is thinking about causing harm to him- or herself, an EMT must do an assessment for suicide risk. What are the signs that a patient may exhibit that could lead to self-harm?

The patient may act depressed, sad, or tearful for extended periods; the patient may voice thoughts of death or taking his or her life; any suicidal gesture must be taken seriously.

There are certain risk factors that are associated with the likelihood of a person committing suicide. Identify them.

Persons over 40 years old; single, widowed, or divorced; alcoholic or depressed; may have a defined lethal plan of action that has been verbalized; an unusual gathering of articles that can cause death, such as the purchase of a gun, large volume of pills, etc.; person may have a previous history of self-destructive behavior or a recent diagnosis of a serious illness; potential problems: the recent loss of a significant loved one, arrest, imprisonment, or the loss of a job.

Describe the function of the right atrium of the heart.

The right atrium receives blood from the veins in the body and the heart; it pumps oxygen-poor blood to the right ventricle.

Describe the function of the left atrium.

The left atrium receives blood from the pulmonary veins located in the lungs; it pumps oxygen-rich blood to the left ventricle.

Describe the function of the right ventricle.

The right ventricle pumps blood to the lungs.

Describe the function of the left ventricle.

The left ventricle pumps blood to the rest of the body.

Describe the function of the valves located within the heart.

Prevent back-flow of blood into the chambers from which it came.

Describe a unique characteristic of the heart muscle itself that is not found in any other muscle in the body.

Contains its own conductive system.

Describe the function of the arteries that are located within the body.

Carry blood from the heart to the rest of the body.

What is the function of the coronary arteries?

To supply blood to the heart muscle.

Describe the major function of the aorta.

The aorta is the major artery originating from the heart and lying in front of the spine in the thoracic and abdominal cavities; it provides blood flow to the rest of the vascular system in the body.

Describe the location and function of the pulmonary artery.

The pulmonary artery originates in the right ventricle; its function is to carry oxygen-poor blood to the lungs.

Describe the location and function of the carotid artery.

The carotid artery is located in the neck; it supplies the head with blood. Pulsation can be palpated on either side of the neck.

Describe the location and function of the femoral artery.

The femoral artery is the major artery of the thigh; it supplies the groin and lower extremities with blood. Pulsation can be palpated in the groin.

Describe the location of the radial artery.

The radial artery provides blood supply to the lower hand; pulsation can be palpated at the wrists on the thumb side.

Describe the location and function of the brachial artery.

The brachial artery is found in the upper arm; it supplies blood to the humerus, the biceps, and triceps; pulsation can be palpated on the inside of the arm between the elbow and shoulder; it is often used when measuring a blood pressure with a blood pressure cuff and a stethoscope.

Describe where you would find the posterior tibial pulse.

On the posterior surface of the medial malleolus.

Identify the location of arterioles within the circulatory system.

Arterioles are the smallest branches of an artery leading to the capillaries.

Describe the location and function of capillaries.

Capillaries are tiny blood vessels that connect arterioles to venules; they are found in all parts of the body; their function is to allow the exchange of nutrients and waste at the cellular level.

What are venules?

Venules are the smallest branches of veins leading to the capillaries.

Identify the function of veins within the cardiovascular system.

Veins are vessels that carry blood back to the heart; the only exception to this rule is the pulmonary vein, which carries oxygenated blood from the lungs back to the heart.

What is the function of the pulmonary vein?

To carry oxygen-rich blood from the lungs to the left atrium for distribution to the rest of the body.

What is the location and function of the vena cava?

The vena cava is broken into two sections: the superior vena cava, which originates in the head, and the inferior vena cava, which collects deoxygenated blood from the rest of the body. The function is to carry blood to the right atrium.

What is the function of red blood cells?

To carry oxygen to the organs of the body and to carry carbon dioxide away from those organs; they give blood its color.

What is the function of white blood cells within the circulatory system?

To act as part of the body's defense against infection.

Describe the function of plasma as it relates to the circulatory system.

The fluid carries the blood cells and nutrients.

Describe the function of platelets within the cardiovascular system.

Platelets are essential for the formation of blood clots.

Describe the meaning of the term "pulse" as it relates to the cardiovascular system and explain where a pulse could be palpated.

The pulse is what is felt when the left ventricle contracts and sends a wave of blood through the arteries; it can be palpated anywhere an artery simultaneously passes near the skin's surface and over a bone.

Identify four primary sites where a peripheral pulse may be palpated.

Radial; brachial; posterior tibial; dorsalis pedis.

Identify two areas in which a central pulse may be palpated.

The carotid artery in the neck or the femoral artery in the groin.

What is meant by the following two terms as they relate to blood pressure: systolic and diastolic.

Systolic is the pressure exerted against the walls of the artery when the left ventricle contracts; diastolic is the pressure exerted against the walls of the artery when the left ventricle is at rest.

An EMT is often presented with a patient who is in shock. Shock is also called hypoperfusion. Identify the meaning of inadequate circulation as it relates to shock.

Shock or inadequate circulation is a state of profound depression of the vital processes of the body. It is characterized by signs and symptoms such as pale, cyanotic, cool, clammy skin; rapid, weak pulse; rapid, shallow breathing; restlessness, anxiety, or mental dullness; nausea and vomiting; reduction in the total blood volume; low or decreasing blood pressure; and subnormal temperatures.

As an EMT, you will often respond to calls for heart problems. Cardiac compromise is often a problem associated with the prehospital setting. List several signs and symptoms of a patient who may be suffering cardiac compromise.

A squeezing feeling in the chest; dull pressure; chest pain commonly radiating down the arms or into the jaw; a sudden onset of sweating; difficulty in breathing; anxiety; irritability; a feeling of impending doom; abnormal pulse rate that may be irregular; abnormal blood pressure; epigastric pain; nausea; vomiting.

Describe the appropriate emergency medical care for a patient who is complaining of having chest discomfort with related heart disease.

Assess baseline vital signs; administer oxygen; encourage the patient to cease all activity; sit the patient in a position of comfort; loosen tight clothing around the waist and neck; begin obtaining a focused assessment.

Identify several things the EMT should question the patient about his or her chest discomfort.

When did the pain start? What caused the pain to begin? Have the patient rate the quality of the pain, 1 being the lowest and 10 being the worst pain ever experienced. Is there any radiation of the pain to the arm, neck, or jaw?

Describe the proper procedure for administering nitroglycerin to a patient complaining of chest pain.

Determine if nitroglycerin has been prescribed to the patient and if it is with the patient; assess the patient's blood pressure to ensure that the systolic pressure is higher than 100 mm Hg; administer one dose and repeat in 3–5 minutes if no relief and if authorized by medical direction, up to a maximum of three doses; reassess vital signs and chest discomfort after each dose; if the patient's blood pressure is less than 100 mm Hg systolic, continue with the elements of the focused assessment and do not administer the nitroglycerin.

Describe the treatment of a patient complaining of chest discomfort who does not have prescribed nitroglycerin.

Give oxygen at 15 L/min via nonrebreathing mask while continuing with the focused assessment; transport promptly; assess the need for advanced life support.

Describe the importance of automated external defibrillation to the EMT-B.

Successful resuscitation of out-of-hospital arrest depends on a series of critical interventions known as the chain of survival; AED is the third element in the chain.

Identify the components of the American Heart Association chain of survival.

Early access; early CPR; early defibrillation; early ACLS.

Describe the rationale for early defibrillation by the EMT-B in the prehospital setting.

Many EMS systems have demonstrated increased survival outcomes of cardiac arrest patients experiencing ventricular fibrillation when early defibrillation is used in the prehospital setting.

Identify a major factor attributed to the increase in survival of cardiac arrest patients in the prehospital setting.

Increased survival of cardiac arrest patients in the prehospital setting is directly attributed to early defibrillation programs that are implemented when all the links in the chain of survival are present.

Identify the features of a fully automated external defibrillator.

Operates without action by EMTs, except to turn the power on; all operations of the unit are contained within the apparatus; no intervention is needed by the EMT.

Describe the features of a semiautomated external defibrillator.

Uses a computer voice synthesizer to advise the EMT of the steps to take based on the AED's analysis of the patient's cardiac rhythm.

Describe how an AED analyzes the cardiac rhythms of a patient.

The AED computer microprocessor evaluates the patient's rhythm and confirms the presence of a rhythm for which a shock is indicated; accuracy of these devices in rhythm analysis has been high, both in detecting rhythms needing shocks and rhythms that do not need shocks; analysis is dependent on properly charged defibrillator batteries.

Indicate two reasons the AED may inappropriately deliver shocks.

Human error; mechanical error.

Identify the proper criteria for attaching the AED to a cardiac patient.

Attach the AED only to unresponsive, pulseless, nonbreathing patients to avoid delivering inappropriate shocks.

When properly using an AED in a cardiac arrest patient, the interruption of CPR is necessary. Identify the criteria for which CPR should be interrupted.

No CPR is performed at times when shocks are delivered; no person should be touching the patient when the rhythm is being analyzed and when shocks are being delivered; cardiac compressions and artificial ventilations are stopped when the rhythm is being analyzed and when more shocks are delivered; defibrillation is more effective than CPR, so stopping CPR during the process is more beneficial to the patient outcome.

How long may CPR be interrupted at any given time?

CPR may be stopped up to 90 seconds if three shocks are necessary; resume CPR only after up to the first three shocks are delivered.

As an EMT, you should be able to describe several advantages of automated external defibrillation. Identify those advantages.

The use of the AED is easier to learn than CPR, however, you must memorize the treatment sequence; EMS delivery system should have all the necessary links in the chain of survival; should be medical direction indicating the proper use of the AED; the first shock should be delivered within 1 minute of the arrival of the unit at the patient's side.

Identify each of the operational steps necessary to begin using the AED in an unconscious, unresponsive patient.

Take infection control precautions en route to the incident; arrive on scene and perform initial assessment to include ABCs; stop CPR if in progress; verify pulselessness and apnea; have your partner resume CPR; attach the AED device; turn on defibrillator power; if the machine has a tape recorder, begin narrative at this time; stop CPR; clear the patient and initiate the analysis of the patient.

Identify the proper procedure the EMT should use if the AED advises shock.

Make sure no one is touching the patient, then push the shock button; immediately begin five cycles of CPR beginning with chest compressions; after five cycles of CPR, reanalyze the patient's rhythm; if the AED advises a shock, push the shock button; if no shock is advised, check for a pulse; if patient has a pulse, check for breathing; if the patient is breathing adequately, give patient oxygen via nonrebreathing mask and transport; if the patient has no pulse, perform five cycles of CPR; after five cycles of CPR, push the analyze button again; if necessary, repeat cycles of CPR, analyzing rhythm, and shock until ALS arrives.

Identify when the patient should be transported, assuming no ALS intervention is available on scene.

The patient should be transported when one of the following occurs: the patient regains a pulse; six shocks are delivered; the machine gives three consecutive messages separated by 1 minute of CPR, but no shock is advised.

After completing an assessment of the ABCs, describe all treatments that proceed defibrillation in the prehospital setting.

Following assessment of the ABCs, defibrillation comes first; do not hook up oxygen or any other procedure that delays analysis of rhythm or defibrillation.

The EMT must be familiar with the AED device used in his or her operational EMS setting. Identify several universal procedures that are important for the EMT-B to understand prior to implementing the use of this device.

All contact with the patient must be avoided during the analysis of the rhythm; state, "Clear the patient" before delivering any shocks; no defibrillator is capable of working without properly functioning batteries; check all batteries at beginning of shift; it is advisable to carry extra batteries.

Identify the age and weight guideline when using the AED.

Automated external defibrillation is not to be used in cardiac arrest in children younger than 12 years of age or under 90 pounds of weight.

True or false: Automated external defibrillators can analyze rhythms when the emergency vehicle is in motion.

False. Automated external defibrillators cannot analyze rhythms when the emergency vehicle is in motion; vehicle must completely stop to analyze rhythm before shocks are ordered—plus, it is not safe to defibrillate in a moving ambulance.

Identify the most frequent reason for AED defibrillator failure.

AED defibrillator failure is most frequently related to improper device maintenance, commonly battery failure; ensure proper battery maintenance and battery replacement schedules.

Does successful completion of AED training in an EMT-B course permit usage of the device without the approval of state laws, rules, and local medical direction authority?

No.

Identify the indications that must be presented to the EMT for him or her to assist a patient in taking the nitroglycerin tablets that are prescribed.

The patient must exhibit signs and symptoms of chest pain and have physician-prescribed sublingual tablets; EMT must have specific authorization by medical direction to assist the patient.

Identify several contraindications in giving a patient a nitroglycerin pill.

Patient should not be given nitroglycerin if he or she: has hypotension or blood pressure below 100 mm Hg systolic; has a head injury; is an infant or child; has already taken the maximum prescribed dose prior to the EMT arrival. (Maximum prescribed dose is three tablets.)

Identify the proper procedure for administering a nitroglycerin tablet to a patient with chest pain.

Obtain order from medical direction either online or off-line; perform a focused cardiac assessment; assess the blood pressure to ensure a systolic of greater than 100 mm Hg; contact medical control if there are no standing orders; ensure that you have the right medication, the right patient, the right route, and the patient is alert and up to taking the product; check the expiration date of the nitroglycerin; question the patient as to last dose, the effects, and ensure that he understands that he is going to take his own medication; ask the patient to lift his tongue and place the tablet or spray dose under the tongue while wearing protective gloves; have the patient keep his mouth closed with the tablet under the tongue without swallowing until it is dissolved and absorbed; recheck the blood pressure within 2 minutes; have the patient rest; provide high-flow oxygen via nonrebreathing mask; prepare for transfer to the local hospital.

Describe the actions of nitroglycerin on the patient after taking the medication.

Decreased the workload of the heart and dilation of blood vessels, which allows for greater blood flow to the heart muscle.

Identify potential side effects of nitroglycerin.

Hypotension; headache; changes in the pulse rate.

Identify the reassessment strategies an EMT should use when administering nitroglycerin to a cardiac patient.

Monitor the blood pressure every 3–5 minutes; ask the patient about effect on pain relief; seek medical direction before administering a second dosage and record all reassessments.

Identify five ways in which a patient may lose heat.

Radiation; convection; conduction; evaporation; breathing.

Identify the condition in which a patient's heat loss exceeds the patient's heat gain.

Hypothermia.

Identify the condition in which a patient's heat gain exceeds the patient's heat loss.

Hyperthermia.

Identify those questions that are important to ask regarding patients suffering from exposure to the environment.

What is the source of their exposure? What particular environment were they in? Have they experienced any loss of consciousness? What effects are they feeling in their body?

Infants and young children are at great risk of generalized hypothermia. What factors increase their risk?

They are small with large surface areas; small muscle mass does not allow adequate shivering in children and none at all in infants; they have less body fat to insulate them from the environment. Younger children need help to protect themselves; they are unable to put on or take off their clothes, which afford them protection in a given environment.

Identify the signs and symptoms of generalized hypothermia and how you would assess a patient experiencing hypothermia.

Cold or cool skin temperature. EMT-B should place the back of the hand between the clothing and the patient's abdomen to assess the general temperature of the patient; patient experiencing a generalized cold emergency will present with cold abdominal skin temperature; a decreasing mental status or motor function directly correlates with a degree of hypothermia.

Identify the respiratory variations that may be seen in both the early and late hypothermic patient.

Patients suffering from early signs of hypothermia may experience rapid breathing; patients suffering from late hypothermia may experience shallow, slow, or even absent breathing.

A patient experiencing hypothermia will show variances in his or her pulse rate depending on the severity of the hypothermia. Describe these changes.

Early hypothermia shows with a rapid pulse rate; late hypothermia shows with a slow, barely palpable, and/or irregular or completely absent pulse rate; the patient may also have a low to absent blood pressure.

Describe the skin of a patient suffering from hypothermia.

May exhibit skin that is red in the early stages of hypothermia and, as hypothermia progresses, skin becomes pale, even cyanotic; skin texture may become stiff and hard.

Identify the appropriate medical care for a patient suffering from generalized hypothermia.

Remove the patient from the environment; protect the patient from further heat loss; place in a warm environment as soon as possible; remove all wet clothing and cover with a blanket; handle the patient extremely gently; avoid rough handling of any kind; do not allow the patient to walk or exert him- or herself in any fashion; administer oxygen if not already done as part of the initial assessment; oxygen administered should be warmed and humidified if possible; assess pulses for 30–45 seconds before starting CPR, as the patient may be bradycardic; the 30 or 45 seconds ensures a good time frame for assessing pulses; if the patient is alert and responding appropriately, actively rewarm the patient with warm blankets, heat packs, or hot water bottles to the groin, axillary, and cervical regions; if possible, turn up the heat in the patient compartment of the ambulance.

Describe the emergency medical treatment for a patient suffering from hypothermia who is unresponsive or not responding appropriately.

Rewarm using warm blankets; turn the heat up in the patient compartment of the ambulance; do not allow the patient to eat or drink stimulants; do not massage the extremities or any localized areas that may have been exposed to the extreme cold.

Conditions occur in which a patient may suffer a localized cold injury to a specific area of the body. Identify those areas of the body that are particularly at risk in extreme cold temperatures.

These areas tend to occur on the extremities and the exposed ears, nose, and face.

Identify signs and symptoms of local cold injuries to the ears, nose, and face.

Skin involved in a localized injury from the cold has lines of clear demarcation; if there is early or superficial injury, there may be blanching of the skin; palpate the skin in which the normal color does not return; check for loss of feeling and sensation in the injured area; skin may remain soft; if rewarmed, tingling sensation may occur as well; in late or deep injuries, a white waxy skin is often noticed; in firm to frozen skin, swelling and blisters may be present; if thawed or partially thawed, the skin may appear flushed with areas of purple and blanching or mottled and cyanotic.

The EMT must be able to provide proper emergency medical care for a patient suffering from localized cold injuries. Identify the proper emergency medical care.

Remove the patient from the environment immediately; protect the cold injured extremity from further injury; administer oxygen if not already done as part of the initial assessment; remove all wet or restrictive clothing; if early or superficial injury, splint the extremity; cover the extremity and do not rub it; do not reexpose to the cold; if it is a late or deep-cold injury, remove all jewelry and cover with dry clothing or dressings; do not break any blisters, rub, or massage the area; apply heat directly to the tissue or attempt to rewarm; do not allow the patient to walk on any of the affected extremities; when an extremely long or delayed transport is inevitable, then active rapid rewarming should be done.

List the proper procedure for active rapid rewarming.

Immerse the affected part in a warm water bath; monitor the water to ensure that it does not cool from the frozen part; continuously stir the water to keep it moving; continue until the part is soft and color and sensation return; dress the area with a dry sterile dressing; if it is a hand or a foot, place a dry sterile dressing between fingers and toes; protect against refreezing of the warmed part; expect the patient to complain of severe pain.

How do high ambient temperatures affect the body's ability to cool itself?

High ambient temperatures reduce the body's ability to lose heat by radiation.

How does high humidity affect the body's ability to cool itself?

High relative humidity reduces the body's ability to lose heat through evaporation.

During exercise and extreme activity, a patient may lose more than 1 liter of sweat per hour. Why may this lead to a heat emergency?

Loss of more than 1 liter of fluid per hour leads to the loss of electrolytes such as sodium, chloride, and other fluids through sweat; may put the patient at risk for dehydration.

Both the elderly and the very young are at high risk for complications related to heat emergencies. Identify reasons for their high risk.

The elderly are at risk for the following reasons: poor thermoregulation; taking medications that affect their temperature regulation; lack of mobility to leave a hot environment. Children, newborns, and infants have poor thermoregulation and are unable to remove their own clothing once placed into a warm environment, which puts them at greater risk for heat emergencies.

Identify several signs and symptoms a patient may exhibit suffering from a heat emergency.

Muscular cramps, weakness or exhaustion, dizziness or faintness; skin may be moist, pale, and cool in temperature; skin may be hot and dry or moist, which is considered a dire emergency; the heart rate may be rapid; patient may experience a rapid alteration in his or her mental status, including unresponsiveness.

Identify the proper emergency medical care for heat emergencies where the patient has warm, moist, pale skin.

Remove the patient from the hot environment and place in a cool environment such as the back of an air-conditioned ambulance; administer oxygen if not already done during the initial assessment; loosen or remove all clothing; cool patient by fanning; place the patient in the supine position with legs slightly elevated; if the patient is responsive and is not nauseated, have the patient drink water; if the patient is unresponsive or is vomiting, place the patient on the left side and transport to the hospital; do not provide the patient with fluids by mouth.

Identify the proper emergency medical care for heat emergencies where the patient has hot, dry, or moist skin.

Remove the patient from the hot environment and place in a cool environment, preferably in the back of an air-conditioned ambulance running on high; remove clothing; administer oxygen if not already done during initial assessment; apply cool packs to the neck, groin, and armpits; keep the skin wet by applying water by sponge or wet towels; fan aggressively; transport immediately.

At any time, the EMT may encounter a water-related emergency. Identify several issues the EMT should consider while dealing with a near drowning or drowning victim.

Ensure the safety of all rescue personnel first and foremost; suspect possible spine injury if a diving accident is involved or unknown; consider length of time in cold-water drowning; any pulseless, nonbreathing patient who has been submerged in cold water should be resuscitated and transported to a medical facility for further evaluation and care.

Identify the appropriate emergency medical care for a patient who is suspected to have suffered a near drowning.

Inline immobilization and removal from water with backboard if spinal injury is suspected and patient is responsive; if there is no suspected spine injury, place the patient on the left side to allow water, vomitus, and secretions to drain from the upper airway; suction as needed; administer oxygen if not already done during the initial assessment; if gastric distention interferes with artificial ventilation, the patient should be placed on the left side with suctioning immediately available; the EMT should place his or her hand over the epigastric region of the abdomen and apply firm pressure to relieve the distention; this procedure should be done only if gastric distention interferes with the ability of the EMT to artificially ventilate the patient effectively.

Identify the appropriate emergency medical care for a patient who has been stung by an insect or bitten by a snake.

If a stinger is present, remove it by scraping the stinger out with the edge of a card or other appropriate device; avoid using tweezers or forceps as these can squeeze venom from the venom sac into the wound; gently wash the area; remove jewelry from the area if possible if swelling begins; place the injection site slightly below the level of the patient's heart so as to slow systemic circulation; do not apply cold to snake bites; consult medical direction regarding constricting band for snakebite; observe for development of signs or symptoms of an allergic reaction and treat as appropriate for an allergic reaction if encountered.

Define the term fetus.

The fetus is the developing baby found in the uterus of the mother.

What is the function of the uterus?

The uterus is the organ in which the fetus grows; it is responsible for labor and expulsion of the infant.

What is the birth canal?

Consists of the vagina and lower part of the uterus through which the fetus is expelled.

What is the function of the placenta?

The placenta is the organ through which the fetus exchanges nourishment and waste products during pregnancy.

Describe the function of the umbilical cord.

The umbilical cord is an extension of the placenta through which the fetus receives nourishment and oxygenated blood while in the uterus.

The fetus is protected within the uterus by a bag of water. Identify that structure and describe its function.

The amniotic sac is the bag of water; the sac surrounds the fetus inside the uterus and protects the fetus while it is developing.

What is the perineum?

The skin area between the vagina and the anus; it is commonly torn during delivery.

Upon arrival, the EMT is presented with a full-term mother with delivery imminent. The mother tells the EMT that she is beginning to crown. What should this mean to the responding EMT?

Crowning is the bulging out of the vagina, which is opening as the fetus's head or presenting part presses against it just prior to birth.

Arriving EMTs question an expectant mother concerning her imminent delivery. She tells them that she has had a "bloody show." What should this term mean to the EMTs?

Bloody show is the mucous and blood that comes out of the vagina as labor begins.

What is the definition of labor?

Labor is defined as the time and process (defined in three or four stages) beginning with the first uterine muscle contraction until delivery of a placenta.

What is the "presenting part"?

The part of the infant/fetus that comes first—usually the head.

Captain Birney is preparing his emergency childbirth kit following a previous run. He is obtaining all equipment necessary to restock the kit. What supplies should he place in his kit?

Surgical scissors; hemostats or cord clamps; umbilical tape or sterilized cord; a bulb syringe for suctioning; clean or sterile towels; 2 x 10 gauze sponges; sterile gloves; a baby blanket; sanitary napkins; a plastic bag.

Describe the emergency medical care for predelivery emergencies such as miscarriage or spontaneous abortion.

Size up the situation to determine the need for advanced life support; perform an initial assessment to include ABCs and vital signs; obtain a history and physical exam and base your treatment on those signs and symptoms; if necessary, apply external vaginal pads; do not pack the vaginal canal, allow fluids to drain; bring any exposed or expelled fetal tissue to the hospital; support the mother.

Describe the emergency medical care for a patient having a seizure during her pregnancy.

Perform an initial assessment to include the ABCs; assess baseline vital signs; treatment is based on those signs and symptoms; transport on her left side.

Describe the emergency medical treatment for vaginal bleeding late in the pregnancy, with or without pain.

Perform an initial assessment to include the ABCs, history, and physical exam; assess baseline vital signs; apply external vaginal pads to control bleeding; do not pack the vaginal canal, allow fluids to drain; transport patient in the left lateral recumbent position, if possible.

Describe the emergency medical care for a pregnant female who may have been involved in a traumatic incident.

Trauma care is the same as for all other patients. Ensure ABCs and ensure cervical spine immobilization if necessary; do an initial assessment to include ABCs and vital signs; obtain a history and physical exam; assess baseline vital signs; treatment should be based on signs and symptoms; transport the patient on the left lateral recumbent position; support with high-flow oxygen on a nonrebreathing mask; ensure C-spine immobilization if applicable.

As an EMT, you respond to a call for an OB. Upon arrival at the 15th floor of an office building, you find a 24-year-old conscious, alert female who appears to be in imminent childbirth. The EMT should ask the following questions to determine whether it is best to transport the expecting mother unless delivery is expected within a few minutes. What questions would you as the EMT ask the expectant mother?

How long have you been pregnant? Are there contractions or is there pain? Is there any bleeding or discharge? Do you have crowning occurring with contractions? What is the frequency and duration of your contractions? Do you feel as if you are having a bowel movement with increasing pressure in the vaginal area? Do you feel the need to push? Is your abdomen rock hard? All of these questions need to be answered to assist you in determining the next course of action.

Several precautions need to be taken by the EMT responding to a call of childbirth. Identify those precautions.

Use body substance isolation, preventing splashing fluids; do not touch the vaginal area except during delivery and when your partner is present; do not let the expectant mother go to the bathroom; do not hold the mother's legs together; recognize your own limitations and transport even if delivery must occur during transport; if delivery is imminent with crowning, contact medical direction for decision to commit to delivery on site; if delivery does not occur within 10 minutes, contact medical direction for permission to transport; support the mother with high-flow oxygen; prepare an obstetrical kit and seek medical direction.

Describe the procedures that you will take as an EMT to deliver a child on scene.

Apply gloves, mask, gown, and eye protection for infection control precautions; have the mother lie with knees drawn up and spread apart; elevate her buttocks with blankets or pillows; clear the sterile field, surrounding the vaginal opening with sterile towels or paper barriers; support mother with oxygen via nasal cannula at 6 L/min; when the infant's head appears at crowning, place fingers on bony part of skull; take care not to touch the fontanel or face; exert very gentle pressure to prevent explosive delivery; use caution to avoid the fontanel; if the amniotic sac does not break or has not broken, use a clamp to puncture the sac and push it away from the infant's head and mouth as they appear; as the infant's head is being born, determine if the umbilical cord is around the infant's neck; slip over the shoulder or clamp, cut, and unwrap; after the infant's head is born, support the head; suction the mouth and the nostrils two or three times; use caution to avoid contact with the back of the mouth; as the torso and full body are born, support the body with both hands; as the feet are born, grasp the feet; wipe blood and mucous from the mouth and nose with a sterile gauze; suction the mouth and nose again, ensuring your grasp of the bulb syringe prior to placing it into the child's nose or mouth; wrap the infant in a warm blanket and place on its side, head slightly lower than the trunk; keep the infant level with the vagina until the cord is cut; assign your partner to monitor the infant and complete initial care; approximately four fingers from the infant, clamp, tie, and cut the umbilical cord between the clamps as pulsations cease; observe for delivery of placenta while preparing mother and infant for transport; when delivered, wrap placenta in towel and place it in plastic bag; transport the placenta to the hospital with the mother; place a sterile pad over the vaginal opening; lower the mother's legs and help her hold them together; record a time of delivery and transport the mother, infant, and placenta to the hospital; ensure supportive oxygen to both the child and the mother.

Following vaginal delivery, vaginal bleeding of up to 500 cc of blood loss is noted. Is this of concern and what should the EMT do about it?

No. 500 cc blood loss is well tolerated by the mother following delivery.

Describe what is considered excessive blood loss following a vaginal delivery. What is the procedure to stop excessive bleeding?

Blood loss in excess of 500 cc. Procedure at this point is to massage the uterus, with fingers fully extended; place on the lower abdomen above the pubis; massage or knead over the area; if bleeding continues, check massage technique and transport immediately, providing high-flow oxygen via nonrebreathing mask and an ongoing assessment.

After stabilizing the mother following delivery, what is the initial care of the newborn?

Position, dry, wipe, and wrap the newborn in a blanket and cover the head; repeat suction of nose and mouth; assess the infant; there should be no central cyanosis; the pulse should be greater than 100 beats per minute; there should be grimace and vigorous crying; there should be good activity and good motion in the extremities; the breathing effort should be normal; the child may be crying; the respiratory rate should be in the 30–35 range; if not breathing, stimulate the newborn by flicking the soles of the feet or rubbing the infant's back.

It is extremely important for the EMT to continually assess the newborn infant's breathing effort. Describe the procedure for this assessment.

The breathing effort is noted by the rise and fall of the child's chest, a lack of, or presence of breath sounds bilaterally; if the breathing effort is slow, shallow, or absent, provide artificial ventilations at a rate of 60 per minute and reassess after 30 seconds; if no improvement, continue artificial ventilation and reassess.

Assessment of the newborn infant's heart rate is extremely important. Describe the procedure for assessing the child's heart rate and emergency treatment for a child with a heart rate less than 100 beats per minute.

The heart rate should be assessed by listening for an apical pulse; if the heart rate is less than 100 beats per minute, provide artificial ventilation at a rate of 60 per minute; reassess after 30 seconds; if no improvement, continue artificial ventilation and reassess; if less than 80 beats per minute and not responding to bag-mask, begin chest compressions; if less than 60 beats/min, start compressions and artificial ventilation.

The EMT must be prepared for abnormal deliveries that occur in a prehospital setting. What is the condition known as a prolapsed cord?

A condition in which the cord presents through the birth canal before delivery of the head; this presents a serious emergency, which endangers the life of the unborn fetus.

Describe the emergency medical care for a patient who presents with a prolapsed cord.

The EMT should perform an initial assessment to include the ABCs and vital signs; support the mother with high-flow oxygen via nonrebreathing mask; obtain a history and physical exam; obtain and assess baseline vital signs; position the mother with the head down or buttocks raised using gravity to lessen pressure in the birth canal; if need be, insert sterile-gloved hand into vaginal canal, pushing the presenting part of the fetus away from the pulsating cord; rapidly transport keeping pressure on presenting part and monitoring pulsation of the cord; contact medical control as soon as possible.

What is meant by breech birth presentation?

Occurs when the buttocks or lower extremities are low in the uterus and will be the first part of the fetus delivered.

Describe the emergency medical care the EMT should provide to the expectant mother when presented with a breech birth.

Assess vital signs and ABCs; provide mother with high-flow oxygen via nonrebreathing mask; immediate and rapid transportation upon recognition; place mother in head-down position with pelvis elevated.

Medic 2 is dispatched to a report of an OB. Upon arrival, they are confronted with a limb presentation. What does this mean to the EMT crew?

Occurs when the limb of the infant protrudes through the birth canal; more commonly a foot when the infant is in the breech position.

What is the appropriate emergency medical care the EMT should provide for a mother when presented with a limb presentation?

Immediate rapid transportation upon recognition; be sure to place the mother on a nonrebreathing mask with high-flow oxygen; place the mother in a head-down position with pelvis elevated.

It is possible for the EMT to be presented with multiple births. What course of action is appropriate when presented with such a case?

In the event of multiple births, be prepared for more than one resuscitation; contact medical control at the earliest opportunity and call for additional personnel to assist in the delivery; transport as soon as possible; deliver each child as presented following standard protocol.

This substance appears to be greenish or brownish yellow rather than clear and it is an indication of possible fetal distress during labor. What is it?

Meconium.

What is the proper emergency medical treatment of a child with evidence of meconium staining?

Do not stimulate before suctioning the oropharynx; gently suction the oropharynx using a bulb syringe; maintain airway and transport as soon as possible.

There are two complications inherent in premature deliveries. Identify them.

Premature infants are always at risk for hypothermia, and usually require resuscitation; resuscitation efforts should be started expediently and continued while en route to the hospital.

Describe the emergency medical care to treat a patient suffering from vaginal trauma.

Trauma to the external genitalia is treated as other bleeding soft tissue injuries. Never pack the vaginal canal; provide oxygen and ongoing patient assessment; assess ABCs; ensure a patent airway; provide high-flow oxygen via nonrebreathing mask; transport as directed to closest hospital.

An occasion may arise in which an EMT is presented with a patient complaining of an alleged sexual assault. This is a criminal assault situation, which requires initial and ongoing assessment and management of psychological care. Describe the necessary emergency medical care the EMT should provide for this victim.

Using universal precautions, assess ABCs and vital signs; complete a sample-focused assessment in a nonjudgmental manner; provide crime-scene protection if necessary; examine the genitalia only if profuse bleeding is present; use 4 x 4s or vaginal pads to assist in bleeding control; do not pack the vaginal canal; use same-sex EMT for care whenever possible to alleviate stress on the victim; discourage the patient from bathing, urinating, or cleaning the wounds; report incident to hospital personnel to ensure law enforcement notification.

Medical Emergencies

1. Each medication has a specific set of contraindications. What is a contraindication?
A. The most common use for that drug
B. Situation in which the drug should be given after consultation with a family member
C. Situations in which a drug should not be used because it may cause harm to the patient
D. Dosages as directed by pharmacy

2. For years, EMTs have been administering a medication in a variety of situations that has helped thousands of patients. That medication is:
A. Oxygen.
B. Epinephrine.
C. Aspirin.
D. Nonaspirin pain reliever.

3. The normal breathing rate for a healthy child at rest is:
A. 10–15 breaths per minute.
B. 15–30 breaths per minute.
C. 30 breaths per minute.
D. Above 30 breaths per minute, not to exceed 45 breaths per minute.

4. What harsh sound should an EMT recognize during breathing that indicates that the patient has an airway obstruction?
A. Stridor
B. Dyspnea
C. Ventrical obstruction
D. Total airway collapse

5. A patient suffering from hypoglycemia may appear to have the following physical symptoms EXCEPT:
A. Being highly agitated and aggressive.
B. An intoxicated appearance.
C. Cold and clammy skin.
D. Staggering, slurred speech.

6. Emergency treatment for a patient having a seizure includes the following EXCEPT:
A. Ensure the airway is patent.
B. Provide supplemental low-flow oxygen.
C. Have suction ready.
D. Transport as soon as possible to a hospital.

7. Which of the following is NOT one of the three criteria that must be met prior to using an epinephrine auto-injector?
A. The patient must be exhibiting signs and symptoms of an allergic reaction.
B. The patient's personal epi-pen medication is prescribed by a physician.
C. Medical direction authorizes its use for this patient.
D. Patient must be seizing and over 14 years old.

8. The correct dosage of activated charcoal for a child is 12.5–25 grams. What is the correct dosage for an adult?
A. 25 grams
B. 25–30 grams
C. 25–50 grams
D. 35–60 grams

9. A number of general factors may influence a person's behavior. Which one of the following is NOT a common cause of behavioral alteration?
A. Situational stresses
B. Medical illnesses
C. Postexercise dyslexia
D. Alcohol or drugs

10. Platelets are essential for which of the following?
A. The formation of blood clots
B. The fluids that carry the blood cells
C. The body's defense against infection
D. Collecting deoxygenated blood and cells

11. The EMT attempts to distinguish between the psychological and physical conditions of a patient. Behaviors caused by physical conditions could be:
A. Sleep apnea.
B. Lack of oxygen or low blood sugar (diabetes).
C. High blood sugar (diabetes) and dyspnea.
D. Alcohol or drug abuse.

12. People who are over 40 years of age, widowed or divorced, alcoholic, or depressed are at high risk for:
A. Mild depression.
B. Suicide.
C. Disassociation syndrome.
D. Psychotic behavior.

13. The cardiac muscle is unique, because it contains something that is not found in any other muscle in the body. What does it contain that makes it unique?
A. Its own conduction system
B. Its own ability to absorb oxygen and store oxygen for an extended time
C. Its ability to function with a low blood and oxygen supply
D. Oxygen-poor blood transferance ability

14. Which artery can be palpated in the groin and is the major artery of the thigh?
A. The femoral artery
B. The location for a pressure point to control bleeding
C. Both A and B are correct.
D. Neither A nor B is correct.

15. ______________ are found in all parts of the body and connect arterioles to venules.
A. Veins
B. Smallest branches of venules
C. Minor arterioles
D. Capillaries

16. Radial, brachial, posterior tibial, and dorsalis pedis are four primary sites where the peripheral pulse may be palpated.
A. True
B. False

17. Systolic pressure is the pressure exerted against the walls of the artery when the left ventricle is at rest.
A. True
B. False

18. Place the following steps in the correct order to outline the proper procedure for administering nitroglycerin to a patient complaining of chest pain.
1. Assess the patient's blood pressure.
2. Reassess vital signs and chest discomfort after each dose.
3. Determine if nitroglycerin has been prescribed to the patient.
4. Assess the patient's blood pressure to ensure that the systolic pressure is higher than 100 mm Hg.

A. 1, 4, 3, 2
B. 1, 3, 2, 4
C. 2, 4, 1, 3
D. 2, 3, 4, 1

19. You have administered nitroglycerin and you are monitoring the blood pressure every 3–5 minutes. Before administering a second dose, you must:
A. Check the pulse rate.
B. Contact medical control (medical direction).
C. Check the expiration date of medication.
D. Have the person stand or sit up.

20. Your patient presents with cool skin temperature and decreasing mental status or motor status. You recognize this condition as hypothermia. In addition to placing the back of your hand on the patient's forehead to get a more accurate general temperature, you use the back of your hand to check:
A. The patient's lower extremities.
B. Between the clothes and skin on the upper back.
C. Between the clothing and the patient's abdomen.
D. On either side of the neck at the carotid artery.

21. How does high humidity affect the body's ability to cool itself?
A. Reduces the body's ability to lose heat through evaporation
B. Increases the body's ability to lose heat through evaporation
C. Reduces the body's temperature, which speeds evaporation
D. Both A and C are correct.

22. You just helped remove a teenager from the deep end of a backyard pool. If there is no suspected spinal injury, place the patient:
A. In the recovery position, on his or her right side.
B. In the supine position.
C. On a backboard.
D. On the left side.

23. The emergency treatment for vaginal bleeding late in pregnancy with or without pain includes the use of vaginal pads to control bleeding. To apply the vaginal pads, you should:
A. Place the pad inside the vaginal walls.
B. Place the pad to allow fluids to drain.
C. Place at least four pads if available and secure tightly.
D. Apply external vaginal pads in the same way as a pressure dressing to control bleeding.

24. You determine that conditions indicate birth is imminent; there is crowning and the contractions are 3 minutes apart. How long should you wait to contact medical control if birth doesn't occur?
A. 5 minutes
B. 10 minutes
C. 15 minutes
D. 20 minutes

25. You have assisted in the delivery of a newborn. If the heart rate is less than ______ beats per minute, provide artificial ventilation at a rate of _______ per minute. Reassess after 30 seconds.
A. 120; 80
B. 100; 80
C. 120; 60
D. 100; 60

Section

5 Trauma

Voices of Experience

Keeping Your Focus

During the summer of 1990, while on duty on the busiest medic rig in the city, my crew was dispatched to a call for an elderly woman with a medical problem. While en route to this incident, we were diverted to an intersection only a few blocks away for an "injured man on the tracks." The first responding engine crew flagged us down on a bridge that went over several sets of railroad tracks. We were directed to go over the side of the bridge and down a steep embankment to the tracks below. I quickly looked over the side of the bridge to try to see what had happened.

What I saw didn't make sense at first; a young man was lying parallel to the tracks and there was a second man kneeling beside him, tightening a belt around his thigh. The first man kept trying to get up, but his left arm and leg didn't move with him. As my partner and I and a visiting paramedic friend of mine headed down the embankment, we passed a couple of fire fighters climbing up the embankment. These were veteran inner-city fire fighters, but their wide-eyed expressions told us that this was bad, affecting even them.

When I got to the train tracks, the patient was several yards in front of me, still trying to stand up. He was having difficulty because his left arm had been severed at mid-bicep and his left leg at mid-thigh. I discovered that moments earlier this young man had been "jumping trains" with his uncle, the guy with the belt.

The patient's bleeding was minimal and his stumps waved around, as if he were giving directions. He was alert and asked us if he was going to be all right. I don't recall any of us answering him. We just looked at each other, and without speaking, managed to communicate to each other that this patient's chance of survival was not good. Between the three of us and the fire fighters, we quickly packaged the patient and moved him up the hill to our medic unit. I remember that I carried one of his severed limbs, but I don't remember which one. Once in the ambulance, all three of us began treating him. We managed the ABCs, administered oxygen, controlled the bleeding, and infused intravenous fluids while en route to the trauma center. The patient talked to us throughout the transport.

The hand-off to the awaiting trauma team went smoothly. The trauma center had a dedicated reimplantation team whose only job was to prepare severed body parts for possible reattachment. As we handed them the patient's severed arm and leg, one of them asked, "OK, where's the rest of his arm?"

We told him, "That's all we found, doc. It's probably halfway to Baltimore by now or feeding the local strays." At that point, we had done all we could to try to save the patient's life; it was up to the trauma team now. We began to do the routine postrun activities: filling out the run report, making up the cot, restocking the ambulance, and discussing the run.

One of my memories from this incident is that despite the difficulty of the situation, we as medics treated this patient without losing focus. Each of us maintained composure and worked efficiently. There was no chicken-little running around. We found out later that the patient lived for a couple of days before succumbing to his wounds. We couldn't help wondering if he was better off.

Part of being a good EMT is having the ability to work under pressure, to maintain your focus, and to recall your training in difficult situations. EMS education strives to develop the skills that will be needed to face medical emergencies by placing students in situations of ever-increasing stress to prepare them for the real-world pressure of emergency care in the streets. If you think your instructor is being tough on you and making it stressful, that is an essential part of your training. Being an EMT is not the right profession for everyone; it requires the ability to overcome pressure in order to give patients the best possible care.

There is an old saying that applies to EMS: You have to keep your head while everyone around you is losing theirs. On this particular day, we had to keep our heads while the patient was losing an arm and a leg.

Captain Robert C. Krause, BS, EMT-P
Toledo Fire and Rescue
Toledo, Ohio

SECTION

5

Trauma

"The things which hurt, instruct."
— Benjamin Franklin

Section 5 covers the following subjects:

Lesson 5–1 Bleeding and Shock

Reviews the cardiovascular system, describes the care of the patient with internal and external bleeding, signs and symptoms of shock (hypoperfusion), and the emergency medical care of shock (hypoperfusion).

Lesson 5–2 Soft-Tissue Injuries

Continues with the information taught in Bleeding and Shock, discussing the anatomy of the skin and the management of soft-tissue injuries and the management of burns. Techniques of dressing and bandaging wounds are also taught in this lesson.

Lesson 5–3 Musculoskeletal Injuries

Reviews the musculoskeletal system before recognition of signs and symptoms of a painful, swollen, deformed extremity. Splinting techniques are also taught in this section.

Lesson 5–4 Injuries to the Head and Spine

Reviews the anatomy of the nervous system and the skeletal system. Injuries to the spine and head, including mechanism of injury, signs and symptoms of injury, and assessment. Emergency medical care, including the use of cervical immobilization devices and short and long backboards will also be discussed and demonstrated by the instructor and students. Other topics include helmet removal and infant and child considerations.

Describe the types of body substance isolation that must be routinely taken to avoid skin and mucous membrane exposure to body fluids.

Eye protection, gloves, gown, mask, and hand washing following each run.

When is bleeding considered serious?

If the patient exhibits signs and symptoms of shock (hypoperfusion), regardless of amount of blood loss.

What is the purpose of splinting?

To prevent motion of bone fragments, bone ends, or angulated joints.

What are some of the secondary complications that can occur as a result of a fracture?

Damage to muscles, nerves, or blood vessels; infection; excessive bleeding.

What are the indications for using a traction splint?

A painful, swollen, deformed mid-thigh with no joint or lower leg injury.

What are some contraindications for the use of a traction splint?

Injury to or close to the knee, hip, or pelvis.

Describe the procedure for applying the traction splint.

Assess PMS distal to the injury; provide manual stabilization and apply traction; adjust the splint to proper size; position splint under injured leg and apply ischial strap, then ankle strap; apply mechanical traction; position and secure support straps; reassess PMS.

Can the MAST suit be used to splint lower extremity fractures?

Yes.

Explain why it is important to identify injuries to the head and spinal column during the physical assessment.

Injuries to the head and spine are extremely serious and may result in severe permanent disability or death if improperly treated or missed in the assessment.

How many bones comprise the spinal column?

33.

What is the main purpose of the bones of the spinal column?

To surround and protect the spinal cord.

Describe the proper emergency medical care for a person with a suspected spinal injury.

Establish and maintain in-line immobilization, place the head in a neutral in-line position; maintain constant manual in-line immobilization until the patient properly secured to a backboard with head immobilization.

True or false: With any head injury, the EMT should suspect a spinal injury.

True.

What is the purpose of long backboard immobilization devices?

To provide stabilization and immobilization to the head, neck, torso, pelvis, and extremities.

Describe how long backboards can be used.

To immobilize patients found in a lying, standing, or sitting position.

How many EMTs are needed to maintain in-line immobilization?

One.

Who directs the EMTs in the movement of the patient?

The EMT at the head.

How many EMTs are needed to safely control the movements of the rest of the body?

One to three other EMTs.

Describe the placement of straps to secure a patient to a backboard.

Immobilize the torso to the board by applying straps across the chest, pelvis, and knees; adjust as necessary.

Describe why it is important to understand the concepts of trauma care?

Trauma is the leading cause of death in the United States for persons between the ages of 1 and 44.

What is the definition of perfusion?

The circulation of blood through an organ structure.

Describe the function of perfusion.

Perfusion delivers oxygen and other nutrients to the cells of all organ systems and removes waste products.

What is hypoperfusion?

The inadequate circulation of blood through an organ.

Large amounts of uncontrolled bleeding can result in shock and death. How much blood loss in an adult can result in shock and death?

The sudden loss of one liter (1000 cc).

What are the three types of bleeding that can be found in a trauma patient?

Arterial; venous; capillary.

Describe the appearance of arterial bleeding.

The blood spurts from the wound; it is bright red because it is rich in oxygen.

Describe the appearance of venous bleeding.

The blood flows in a steady stream; it is dark red and low in oxygen content.

Describe capillary bleeding.

The blood oozes from a capillary; it is dark red in color.

Of all the potential types of bleeding the EMT may encounter, which is the most difficult to control?

Arterial bleeding, because of the pressure at which arteries bleed.

This type of bleeding often clots spontaneously. Which is it?

Capillary bleeding.

Describe the proper emergency medical care for controlling external bleeding.

Apply fingertip pressure directly on the point of bleeding using a sterile/clean dressing; elevation of a bleeding extremity may be used secondary to and in conjunction with direct pressure.

If direct pressure and elevation fail to control excessive bleeding, what additional technique can be employed?

Pressure points may be used in the upper arm and lower extremities.

Identify three methods to control external bleeding if direct pressure, elevation, and pressure points are unsuccessful.

To control external bleeding the use of splints, pressure splints, or pneumatic counterpressure devices may be employed.

True or false: A tourniquet can be used as a primary method of bleeding control in most situations to free the EMT to treat other injuries.

False.

Describe the rationale for not using a tourniquet in the prehospital environment.

Application of a tourniquet can cause permanent damage to nerves, muscles, and blood vessels, resulting in the loss of the extremity.

Describe the proper size of a tourniquet.

Use a bandage 4 inches wide and 6 to 8 layers deep.

Identify another piece of equipment that may be used as a tourniquet.

A continuously inflated blood pressure cuff may be used as a tourniquet until bleeding stops.

Patients who have experienced a skull fracture may have bleeding from what two areas?

Bleeding from the nose or ears may occur as a result of a skull fracture.

Define epistaxis.

The medical term for nosebleed.

Describe the emergency medical care for a patient with epistaxis.

Place the patient in the sitting position leaning forward; apply direct pressure by pinching the fleshy portion of the nostrils together.

Why is it necessary for an EMT to recognize a patient who may be suffering from internal bleeding?

Internal bleeding can result in severe blood loss with resultant shock and subsequent death.

At the scene of a traumatic incident an EMT should survey the scene to determine the mechanism of injury. Explain why this is important.

Suspicion and severity of internal bleeding should be based on the mechanism of injury and clinical signs and symptoms.

The EMT must be able to recognize the basic signs and symptoms associated with internal bleeding. What are they?
Pain, tenderness, swelling, or discoloration of a suspected site of injury; altered mental status, restlessness; tender, rigid, and/or distended abdomen.

Trauma patients develop shock from the loss of blood from both external and internal sites. Identify this type of shock.
Shock resulting from fluid and volume loss is referred to as hypovolemic shock or hemorrhagic shock.

Explain why the EMT should not rely on blood pressure as an indicator of a child's or an infant's circulatory status.
Infants and children can maintain their blood pressure until they have lost up to half of their blood volume; by the time their blood pressure drops, they are close to death.

What is the proper height to elevate the legs of a patient in shock?
Approximately 8 to 12 inches.

When during the patient survey should the EMT identify bleeding and shock?
During the initial patient assessment, after securing the scene and ensuring personal safety.

Describe when, during initial assessment, the EMT is to control arterial bleeding.
Control of arterial or excessive venous bleeding will be done immediately on identification, after airway and breathing.

What condition can lead to hypoperfusion (shock)?
Bleeding that is uncontrolled or excessive.

If shock is left uncontrolled, what physiological conditions will result?
Shock will lead to inadequate tissue perfusion and eventual cell and organ death.

List the layers of the skin.
Dermis; epidermis; subcutaneous.

Describe the emergency medical care of the patient with an open soft-tissue injury.
Control excessive bleeding; apply a sterile or clean dressing; secure with a bandage; minimize movement; splint if a fracture is suspected; assess PMS.

State the major functions of the skin.
Protection; water balance; temperature regulation; excretion; shock (impact) absorption.

Identify the types of open soft-tissue injuries.
Abrasion; incisions; lacerations; punctures; avulsion; amputation; crush injury.

Identify the types of closed soft-tissue injuries.
Contusions; internal lacerations and punctures; crush injuries and ruptures.

Describe the purpose of a bandage.
Used to hold a dressing in place.

Describe the emergency medical care of a patient with an impaled object.
Do not remove the impaled object but expose the area of injury; control active bleeding with direct pressure; stabilize the impaled object with a bulky dressing; treat for shock; administer high-flow oxygen with a nonrebreathing mask; calm and reassure the patient; carefully transport the patient as soon as possible.

List the classifications of burns.
Superficial; partial thickness; full thickness.

Describe the emergency medical care for a chemical burn.
Remove the patient from the source of the chemical; assess ABCs; flush effected area with large amounts of water for at least 20 minutes.

A man at home was working on his electrical panel when he contacted a live wire. He received an electrical shock from a 220V line. What is the proper emergency medical care for this patient?
Ensure the scene is safe to enter; it may be necessary to remove the patient from the source of the electric current; assess and control the ABCs, provide BCLS as needed; care for spinal injuries, head injuries, and fractures; assess burn injury; cool the burn area; apply dry sterile dressing; administer high-flow oxygen; and transport as soon as possible.

What factors are used to determine the severity of a burn?

Source of the burn, thermal, chemical, electrical, etc.; area of the body affected; extent and degree of burn; the patient's age and other medical conditions.

Describe the emergency medical care of a patient with a thermal burn.

Stop the burning process; wet down, smother, or remove the clothing; ensure an open airway; assess ABCs; assess airway injury; provide high-flow oxygen; treat for shock; do not pull burnt clothing from burned skin area; wrap with a dry dressing if burn is greater than 9%, moist if less than 9%.

Bob and Guy are checking out their gear on Rescue 13. They are dispatched to a reported "man hit by a train." Upon arrival, Bob looks over the train bridge to see a man lying next to the train track, with bystanders nearby who say he tried to jump on the train and fell off. As Guy and Bob reach the patient, they see that both his left arm and left leg have been amputated and are lying a few yards away. Describe the proper emergency medical treatment for this patient.

Ensure the scene is safe; assess ABCs and control the active bleeding with direct pressure, using pressure points if needed; wrap the injured limbs with sterile pressure dressings to assist in bleeding control; provide high-flow oxygen; wrap or bag the amputated parts in plastic and label; transport the parts with the patient; keep the amputated parts cool; DO NOT immerse in water or saline; reassess for any secondary injuries, treat for shock, and transport as soon as possible, preferably to a trauma center.

What is the definition of evisceration?

Open wound of the abdomen that is so large that organs protrude through the wound opening.

What is the proper emergency medical care for a patient with an evisceration?

Provide high-flow oxygen, treat for shock, do not touch or attempt to replace the protruding organ; cover the exposed area with a dressing soaked with sterile saline; cover entire area with a plastic wrap and secure dressing in place with tape; apply a thick dressing over this area to help prevent heat loss; reassess patient; transport as soon as possible.

True or false: Any partial thickness burn of 10% to 20% is considered a moderate burn in a child.

True.

What are the three functions of dressings and bandages?

Stop bleeding; protect the wound from further damage; prevent further contamination and infection.

Identify four types of dressings that are available to the EMT.

Universal dressing 5 x 9; 4 x 4-inch gauze pads; adhesive-type dressings; occlusive.

Describe the characteristics of a superficial burn.

Reddening of the skin and perhaps some swelling; patient will complain of pain at the burn site; the burn will heal of its own accord, without scarring.

The EMT should be able to assess burn injuries using the Rule of Nines. Using that rule, calculate the following injury in percent of body area burned: adult, head and neck.

9%.

The EMT should be able to assess burn injuries using the Rule of Nines. Using that rule, calculate the following injury in percent of body area burned: adult, each upper extremity.

9%.

The EMT should be able to assess burn injuries using the Rule of Nines. Using that rule, calculate the following injury in percent of body area burned: adult, anterior trunk.

18%.

The EMT should be able to assess burn injuries using the Rule of Nines. Using that rule, calculate the following injury in percent of body area burned: adult, posterior trunk.

18%.

The EMT should be able to assess burn injuries using the Rule of Nines. Using that rule, calculate the following injury in percent of body area burned: adult, each lower extremity.

18%.

The EMT should be able to assess burn injuries using the Rule of Nines. Using that rule, calculate the following injury in percent of body area burned: adult, genitalia.

1%.

The EMT should be able to assess burn injuries using the Rule of Nines. Using that rule, calculate the following injury in percent of body area burned: child, head and neck.

18%.

The EMT should be able to assess burn injuries using the Rule of Nines. Using that rule, calculate the following injury in percent of body area burned: child, each lower extremity.

14%.

The EMT should be able to assess burn injuries using the Rule of Nines. Using that rule, calculate the following injury in percent of body area burned: child, anterior and posterior trunk.

36%.

The EMT should be able to assess burn injuries using the Rule of Nines. Using that rule, calculate the following injury in percent of body area burned: child, each upper extremity.

9%.

Describe the characteristics of a full thickness burn.

The burn extends through all the dermal layers and may involve subcutaneous layers, muscle, bone, or organs; skin becomes dry and leathery and may appear white, dark brown, or charred; little or no pain; may be pain at the periphery of the burned area.

Describe the proper emergency medical care for a patient suffering from a burn injury.

Stop the burning process, initially with water or saline; remove smoldering clothing and jewelry; continually monitor the airway for evidence of closure; prevent further contamination by covering the area with a dry sterile dressing; do not apply any type of ointment, lotion, or antiseptic; do not break blisters; transport as soon as possible.

What is the natural response to bleeding?

Blood vessel contractions and clotting.

What may interfere with effective blood clotting?

Serious injury may prevent effective clotting from occurring.

Describe how you would control bleeding in a large, gaping wound.

Large gaping wounds may require packing with sterile gauze and direct hand pressure if direct fingertip pressure fails to control bleeding; if bleeding does not stop, remove dressing and assess for bleeding point to apply direct pressure; if diffuse bleeding is discovered, apply additional pressure.

Why should a rope, wire, or belt not be used as a tourniquet?

Because they may cut into the skin and underlying tissue.

What procedures should you follow once a tourniquet is in place?

Do not remove or loosen the tourniquet once it is applied unless directed to do so by medical direction; leave the tourniquet in open view.

What are some of the potential causes of epistaxis?

Injured skull; facial trauma; digital trauma (nose picking); sinusitis and other upper respiratory tract infections; hypertension; coagulation disorders.

What is the most serious possible cause of bleeding from the ears or nose?

Bleeding from the ears or nose may occur because of a skull fracture. If the bleeding is the result of trauma, do not attempt to stop the blood flow; collect blood with a loose dressing, which may also limit exposure to sources of infection.

On what basis can you judge the severity of internal bleeding?

Suspicion and severity of internal bleeding should be based on the mechanism of injury and clinical signs and symptoms.

What kinds of injuries often result in blunt trauma?

Falls; motorcycle crashes; pedestrian impacts; automobile collisions; blast injuries.

What signs should you be looking for if you suspect internal injuries as the result of blunt trauma?

Look for evidence of contusions, abrasions, deformity, impact marks, and swelling.

What signs and symptoms of internal bleeding should you be looking for in the case of penetrating trauma?

Pain, tenderness, swelling, or discoloration of suspected site of injury; bleeding from the mouth, rectum, vagina, or other orifice; vomiting bright red blood or dark, coffee ground–colored blood; dark, tarry stools or stools with bright-red blood; tender, rigid, and/or distended abdomen.

Describe the late signs and symptoms of hypovolemic shock (hypoperfusion).

Anxiety, restlessness, combativeness, or altered mental status; weakness, faintness, or dizziness; thirst; shallow, rapid breathing; rapid, weak pulse; pale, cool, clammy skin; capillary refill greater than 2 seconds—infant and child patients only; dropping blood pressure (late sign); dilated pupils that are sluggish to respond; nausea and vomiting.

Describe the emergency medical care for a patient in shock.

Body substance isolation; maintain airway/artificial ventilation; administer oxygen if not already done during the initial assessment; if bleeding is suspected in extremity, control bleeding by direct pressure and application of a splint; transport immediately.

How does shock affect the body on a cellular level?

Shock (hypoperfusion) results in inadequate perfusion of cells with oxygen and nutrients and inadequate removal of metabolic waste products. Cell and organ malfunction and death can result from shock (hypoperfusion); prompt recognition and treatment is vital to patient survival.

How does shock affect peripheral perfusion?

Peripheral perfusion is drastically reduced due to the reduction in circulating blood volume.

What causes hypovolemic or hemorrhagic shock?

Trauma patients develop shock (hypoperfusion) from the loss of blood from both internal and external sites.

How does shock affect blood pressure and pulse?

Early sign: increased pulse rate—weak and thready; late sign: decreased blood pressure.

How does shock affect respiration?

Increased breathing rate; respirations are shallow, labored, and irregular.

Describe a contusion.

Epidermis remains intact; cells are damaged; blood vessels torn in the dermis; blood accumulation causes discoloration

Describe a hematoma.

Collection of blood beneath the skin. Larger amount of tissue damage as compared to contusion; larger vessels are damaged; may lose one or more liters of blood.

Describe a crush injury

The result of a crushing force applied to the body; can cause internal organ rupture. Internal bleeding may be severe with shock.

Describe an abrasion.

Shearing forces that damage the outermost layer of skin; painful injury, even though superficial; no or very little oozing of blood.

Describe a laceration.

Break in skin of varying depth; may be linear (regular) or stellate (irregular) and occur in isolation or together with other types of soft-tissue injury; caused by forceful impact with sharp object; bleeding may be severe.

Describe an avulsion.

A flap or flaps of skin or tissue torn loose or pulled completely off.

Describe a penetration/puncture injury.

Caused by sharp, pointed object; may be no external bleeding; internal bleeding may be severe; exit wound may be present.

Describe an amputation.

Involves separation of the extremities and other body parts; massive bleeding may be present or bleeding may be limited.

Describe crush injuries.

Damage to soft tissue and internal organs; may cause painful, swollen, deformed extremities. External bleeding may be minimal or absent; internal bleeding may be severe.

Describe basic emergency medical care of open soft-tissue injuries.

Utilize appropriate body substance isolation; maintain proper airway/artificial ventilation/oxygenation; expose the wound; control the bleeding; prevent further contamination; apply dry sterile dressing to the wound; bandage securely in place; keep the patient calm and quiet; treat for shock (hypoperfusion) if signs and symptoms are present.

Describe any special considerations in the treatment of a chest injury.

Apply an occlusive dressing to open wound; administer oxygen if not already done; position of comfort if no spinal injury suspected.

In addition to treating for shock, what special concern does an amputation injury present?

Concern for reattachment.

True or false: An amputated part should not be transported with the patient.

False. Transport the amputated part with the patient.

True or false: You should complete a partial amputation for ease of treatment and transport.

False. Do not complete partial amputations; immobilize to prevent further injury.

What serious complication may result from a large open-neck injury?

May cause air embolism.

What special measures can you take in treating a large, open-neck injury?

Cover with an occlusive dressing; compress carotid artery only if necessary to control bleeding.

In addition to the Rule of Nines, how can you estimate the percentage of body area burned?

The size of the patient's hand is equal to 1%.

In addition to severity and extent, what other factors must be considered in determining the seriousness of a burn?

Preexisting medical conditions; age of the patient (less than 5 years of age or greater than 55 years of age).

Describe the criteria for classifying a burn as critical.

Full thickness burns involving the hands, feet, face, or genitalia; burns associated with respiratory injury; full thickness burns covering more than 10% of the body surface; partial thickness burns covering more than 30% of the body surface area; burns complicated by painful, swollen, deformed extremity; moderate burns in young children or elderly patients; burns encompassing any body part, e.g., arm, leg, or chest.

Describe the criteria for classifying a burn as moderate.

Full thickness burns of 2% to 10% of the body surface area excluding hands, feet, face, genitalia, and upper airway; partial thickness burns of 15% to 30% of the body surface area; superficial burns of greater than 50% body surface area.

Describe the criteria for classifying a burn as minor.

Full thickness burns of less than 2% of the body surface area; partial thickness burns of less than 15% of the body surface area.

What is considered a critical burn in a child?

Any full thickness burn or partial thickness burn greater than 20%; burn involving the hands, feet, face, airway, or genitalia.

What is considered a moderate burn in a child?
Any partial thickness burn of 10% to 20%.

What is considered a minor burn in a child?
Any partial thickness burn less than 10%.

True or false: Always immediately remove a patient from the electrical source when treating victim of electrocution.
False. Do not attempt to remove patient from the electrical source unless trained to do so. If the patient is still in contact with the electrical source or you are unsure, do not touch the patient.

True or false: Electrical shock injuries are often more severe than external indications.
True.

What three types of force are often involved in skeletal injury?
Direct; indirect; twisting.

What are the two main classifications of bone or joint injuries?
Open: break in the continuity of the skin; closed: no break in the continuity of the skin.

What are the signs and symptoms of bone or joint injury?
Deformity or angulation; pain and tenderness; grating; swelling; bruising (discoloration); exposed bone ends; joint locked into position.

Describe general emergency medical care of bone or joint injuries.
Body substance isolation; administer oxygen if not already done and indicated; after life threats have been controlled, splint injuries in preparation for transport; application of cold pack to area of painful, swollen, deformed extremity to reduce swelling; elevate the extremity; splint.

Describe the general rules of splinting.
Assess pulse, motor, and sensation distal to the injury prior to and following splint application and record findings; immobilize joint above and below the injury; remove or cut away clothing; cover open wounds with a sterile dressing; if there is a severe deformity or the distal extremity is cyanotic or lacks pulses, align with gentle traction before splinting; do not intentionally replace the protruding bones; pad each splint to prevent pressure and discomfort to the patient; splint the patient before moving when feasible and no life threats; when in doubt, splint the injury when feasible and no life threats; if patient has signs of shock (hypoperfusion), align in normal anatomical position and transport (total body immobilization; example: Backboard takes care of all immobilization on emergency basis).

What are the general categories of splints?
Rigid; traction; pneumatic (air, vacuum); improvised; pillow; pneumatic antishock garment (as a splint).

What are some of the hazards of improper splinting?
Compression of nerves, tissues, and blood vessels from the splint; delay in transport of a patient with life-threatening injury; splint applied too tight on the extremity, reducing distal circulation; aggravation of the bone or joint injury; cause or aggravate tissue, nerve and vessel, or muscle damage from excessive bone or joint movement.

Describe the general procedures for long-bone splinting.
Body substance isolation; apply manual stabilization; assess pulse, motor, and sensory function; measure splint. If there is a severe deformity or the distal extremity is cyanotic or lacks pulses, align with gentle traction before splinting; apply splint, immobilizing the bone and joint above and below the injury; secure entire injured extremity immobilize hand/foot in position of function; reassess pulse, motor, and sensation after application of splint, and record.

Describe the general procedures for splinting a joint injury.
Body substance isolation; apply manual stabilization; assess pulse, motor, and sensory function; align with gentle traction if distal extremity is cyanotic or lacks pulses and no resistance is met; immobilize the site of injury; immobilize bone above and below the site of injury; reassess pulse, motor, and sensation after application of splint and record.

What are the contraindications of the use of a traction splint?

Injury is close to the knee; injury to the knee or hip; injured pelvis; partial amputation or avulsion with bone separation; distal limb connected only by marginal tissue; traction would risk separation; lower leg or ankle injury.

What are the four mechanisms of injuries to the spine?

Compression; excessive flexion, extension, and rotation; lateral bending; distraction.

What mechanisms lead to compression injuries?

Falls; diving accidents; motor vehicle accidents.

What is distraction?

Pulling apart of the spine, such as in hangings.

What mechanisms would cause you to maintain a high index of suspicion for spinal injury?

Motor vehicle crashes; pedestrian-vehicle collisions; falls; blunt trauma; penetrating trauma to head, neck, or torso; motorcycle crashes; hangings; diving accidents; unconscious trauma victims.

True or false: The ability to walk, move the extremities, feel sensation, or lack of pain to the spinal column can rule out the possibility of spinal column or cord damage.

False.

True or false: Do not ask the patient to move to try to elicit a pain response.

True.

True or false: Do not move the patient to test for a pain response.

True.

True or false: While asking questions, tell the patient not to move.

True.

List the signs and symptoms of spinal injury.

Pain independent of movement or palpation along spinal column or lower legs, may be intermittent; obvious deformity of the spine upon palpation; numbness, weakness, or tingling in the extremities; loss of sensation or paralysis below the suspected level of injury; loss of sensation or paralysis in the upper or lower extremities; incontinence.

Certain soft-tissue injuries are associated with spinal trauma. What are they and what areas of the spine are associated with them?

Head and neck—cervical spine; shoulders, back, or abdomen—thoracic, lumbar; lower extremities—lumbar, sacral.

What questions should you ask the victim with possible spinal injury?

Does your neck or back hurt? What happened? Where does it hurt? Can you move your hands and feet? Can you feel me touching your fingers? Can you feel me touching your toes?

Describe the focused physical exam of a responsive patient with possible spinal injury.

Inspect for contusions, deformities, lacerations, punctures, penetrations, and swelling; palpate for areas of tenderness or deformity; assess equality of strength of extremities—hand grip or gently push feet against hands.

Describe the focused physical exam of an unresponsive patient with possible spinal injury.

Inspect for contusions, deformities, lacerations, punctures/penetrations, and swelling; palpate for areas of tenderness or deformity; obtain information from others at the scene to determine information relevant to mechanism of injury or patient mental status prior to the EMT's arrival.

What are the major life-threatening complications of spinal injury?

Inadequate breathing effort and paralysis.

Describe the emergency medical care for a patient with suspected spinal injury.

Body substance isolation; establish and maintain in-line immobilization; place the head in a neutral in-line position unless the patient complains of pain or the head is not easily moved into position; place head in alignment with spine; maintain constant manual in-line immobilization until the patient is properly secured to a backboard with the head immobilized; perform initial assessment; whenever possible, airway control must be done with in-line immobilization; whenever possible, artificial ventilation must be done with in-line immobilization; assess pulse, motor, and sensation in all extremities; assess the cervical region and neck; apply a rigid, cervical immobilization device; properly size the cervical immobilization device. If it doesn't fit, use a rolled towel and tape to the board and have rescuer hold the head manually—an improperly fitted immobilization device will do more harm than good; immobilize.

How should you immobilize a patient found in a lying position?

With a long spine board; hold in-line immobilization using a slide, proper lift, log roll, or scoop stretcher to limit movement to the minimum possible. Which method to use must be decided based on the situation, scene, and available resources.

How should you immobilize a patient found in a sitting position?

If the patient is found in a sitting position in a chair, immobilize with a short spine immobilization device. Exception: If the patient must be removed urgently because of injuries, the need to gain access to others, or dangers at the scene, he or she must then be lowered directly onto a long backboard and removed with manual immobilization provided.

How should you immobilize a patient found in a standing position?

With a long spine board.

What is unique to scalp injuries?

Very vascular; may bleed more than expected.

How do you control scalp bleeding?

With direct pressure.

What are the causes of a nontraumatic brain injury?

May occur due to clots or hemorrhaging.

What are the signs and symptoms of a skull injury?

Mechanism of trauma; contusions, lacerations, and hematomas to the scalp; deformity to the skull; blood or fluid (cerebrospinal fluid) leakage from the ears or nose; bruising (discoloration) around the eyes; bruising (discoloration) behind the ears (mastoid process).

What are the signs and symptoms of a traumatic head injury?

Altered or decreasing mental status; confusion, disorientation, or repetitive questioning; if the patient is conscious—deteriorating mental status; unresponsiveness; irregular breathing pattern; contusions, lacerations, and hematomas to the scalp; deformity to the skull; blood or fluid (cerebrospinal fluid) leakage from the ears and nose; bruising (discoloration) around the eyes; bruising (discoloration) behind the ears (mastoid process); neurologic disability; nausea and/or vomiting; unequal pupil size with altered mental status; seizure activity may be seen.

What are the signs and symptoms of an open head injury?

Contusions, lacerations, and hematomas to the scalp; deformity to the skull; soft area or depression upon palpation; exposed brain tissue; bleeding from the open bone injury; blood or fluid (cerebrospinal fluid) leakage from the ears and nose; bruising (discoloration) around the eyes; bruising (discoloration) behind the ears (mastoid process); nausea and/or vomiting.

What is the emergency medical care of a head injury?

Body substance isolation; maintain airway/artificial ventilation/oxygenation; initial assessment with spinal immobilization should be done on scene with a complete detailed physical exam en route; with any head injury, the EMT must suspect spinal injury; immobilize the spine; closely monitor the airway, breathing, pulse, and mental status for deterioration; control bleeding; do not apply pressure to an open or depressed skull injury; dress and bandage open wound as indicated in the treatment of soft-tissue injuries. If a medical injury or nontraumatic injury exists, place patient on the left side; be prepared for changes in patient condition; immediately transport the patient.

What are the indications for spinal immobilization?

Any suspected injury to the spine based on mechanism of injury, history, or signs and symptoms.

What are some common rules regarding cervical immobilization devices?

Use in conjunction with short and long backboards; improperly sized immobilization device has a potential for further injury; do not obstruct the airway with the placement of a cervical immobilization device; if it doesn't fit, use a rolled towel and tape to the board and manually support the head; an improperly fit device will do more harm than good.

What precautions should you take in regard to cervical immobilization devices?

Cervical immobilization devices alone do not provide adequate in-line immobilization; manual immobilization must always be used with a cervical immobilization device until the head is secured to a board.

What are the two types of short backboards?

Vest-type devices; rigid short boards.

What is the purpose of the short backboard?

Provides stabilization and immobilization to the head, neck, and torso; used to immobilize noncritical sitting patients with suspected spinal injuries.

Describe the general procedures for applying a short backboard.

Start manual in-line immobilization; assess pulses, motor, and sensory function in all extremities; assess the cervical area; apply a cervical immobilization device; position short board immobilization device behind the patient; secure the device to the patient's torso; evaluate torso and groin fixation and adjust as necessary without excessive movement of the patient; evaluate and pad behind the patient's head as necessary to maintain neutral in-line immobilization; secure patient's head to the device; release manual immobilization of head; rotate or lift the patient to the long spine board; immobilize patient to long spine board; reassess pulses, motor, and sensory function in all extremities.

Describe the function of the long backboard.

Provides stabilization and immobilization to the head, neck and torso, pelvis, and extremities; used to immobilize patients found in a lying, standing, or sitting position; sometimes used in conjunction with short backboards.

Describe the general procedures for applying the long backboard.

Start manual in-line immobilization; assess pulses, motor, and sensory function in all extremities; assess the cervical area; apply a cervical immobilization device; position the device; move the patient onto the device by log roll, suitable lift or slide, or scoop stretcher. A log roll is: One EMT must maintain in-line immobilization; EMT at the head directs the movement of the patient; one to three other EMTs control the movement of the rest of the body; quickly assess posterior body if not already done in initial assessment; position the long spine board under the patient; roll patient onto the board at the command of the EMT holding in-line immobilization; pad voids between the patient and the board; immobilize torso to the board by applying straps across the chest and pelvis and adjust as needed; immobilize the patient's head to the board; fasten legs, proximal to and distal to the knees; reassess pulses, motor, and sensation, and record.

What are the indications for rapid extrication?

Unsafe scene or unstable patient condition warrants immediate movement and transport; patient blocks EMT access to another, more seriously injured, patient; based on time and patient, not EMT's preference.

What are the special assessment needs for patients wearing helmets?

Airway and breathing; fit of the helmet and patient's movement within the helmet; ability to gain access to airway and breathing.

What are the indications for leaving the helmet in place?

Good fit with little or no movement of the patient's head within the helmet; no impending airway or breathing problems; removal would cause further injury to the patient; proper spinal immobilization could be performed with helmet in place; no interference with the EMT's ability to assess and reassess airway and breathing.

What are the indications for removing a helmet?

Inability to assess and/or reassess airway and breathing; restriction of adequate management of the airway or breathing; improperly fitted helmet allowing for excessive patient head movement within the helmet; proper spinal immobilization cannot be performed due to helmet; cardiac arrest.

What are the general rules for the removal of a helmet?

Depends on the actual type of helmet worn by the patient; take eyeglasses off before removal of the helmet; one EMT stabilizes helmet by placing hands on each side of the helmet with the fingers on the mandible to prevent movement; second EMT loosens the strap; second EMT places one hand on the mandible at the angle of the jaw and the other hand posteriorly at the occipital region; EMT holding the helmet pulls the sides of the helmet apart and gently slips the helmet halfway off the patient's head and stops; EMT maintaining stabilization of the neck repositions, slides the posterior hand superiorly to secure the head from falling back after complete helmet removal; helmet is removed completely; EMT then can proceed with spinal immobilization as indicated in the spinal immobilization section.

What special measures should be taken when immobilizing infants or children?

Pad from the shoulders to the heels of the infant or child, if necessary, to maintain neutral immobilization; properly size the cervical immobilization device; if it doesn't fit, use a rolled towel and tape to the board and manually support head; an improperly fitted immobilization device will do more harm than good.

Trauma

1. When an EMT applies a traction splint, a number of procedures must be followed in a sequence. Place these four procedures in the correct order.
1. Position the traction splint under the injured leg.
2. Assess PMS distal to the injury.
3. Adjust the splint to the proper size.
4. Provide manual stabilization and apply traction.

A. 1, 2, 3, 4
B. 2, 3, 4, 1
C. 4, 2, 3, 1
D. 2, 4, 3, 1

2. Which of the following is not proper medical care for a person with a suspected spinal injury?
A. Establish and maintain in-line immobilization.
B. Place the patient's head in a neutral in-line position.
C. Apply and maintain mechanical traction.
D. Properly secure the patient to a backboard with head immobilization.

3. Perfusion is the circulation of blood through an organ structure. What is the current function of perfusion?
A. Delivers oxygen and other nutrients to the cells of all organ systems
B. Carries waste products to the kidneys
C. Delivers oxygen to the veins
D. Carries plasma to the body's hollow organs

4. Of all the potential types of bleeding the EMT may encounter, which is the most difficult to control?
A. Venous
B. Capillary
C. Oozing
D. Arterial

5. The proper size of a tourniquet is:
A. 2 inches wide.
B. 4 inches wide.
C. 6 inches wide.
D. 8 inches wide.

6. It is necessary for an EMT to recognize a patient who may be suffering from internal bleeding, because severe blood loss may result in shock and subsequent death.
A. True
B. False

7. Pain, tenderness, swelling, or discoloration of a suspected site of injury and altered mental status, restlessness, tenderness, rigidity, and distention are signs and symptoms of:
A. Circulatory compromise.
B. External bleeding.
C. Late stage of cardiogenic shock.
D. Internal bleeding.

8. What is the final stage of shock if left uncontrolled?
A. Hyperventilation
B. Cell and organ death
C. Hyperactivity
D. Elevated blood pressure and stroke

9. Protection, water balance, temperature regulation, excretion, and shock (impact) absorption, are major functions of:
   A. Muscles and ligaments.
   B. The heart.
   C. Skin.
   D. The brain.

10. Which of the following is not a type of closed soft-tissue injury?
   A. Rupture
   B. Contusion
   C. Avulsion
   D. Internal laceration

11. For a patient with a chemical burn, the EMT will provide care by removing the patient from the source of the chemical, assessing the ABCs, and flushing the affected area with:
   A. Hot water for 5–10 minutes.
   B. Cold water for 5–10 minutes.
   C. Large amounts of water for 10 minutes.
   D. Large amounts of water for at least 20 minutes.

12. An electrician is working in a home and he comes in contact with a live wire. He received an electric shock from a 220V line. What is the first thing you do when you arrive on scene?
   A. Cut the power source.
   B. Ensure that the scene is safe to enter.
   C. Establish responsiveness.
   D. Call the power company.

13. Which of the following is not a factor used in determining the severity of a burn?
   A. The patient's age and other medical conditions
   B. Area of the body affected
   C. Vital signs
   D. Source of the burn

14. Based on the Rule of Nines for burns, what percentage would be considered a moderate burn in a child for any partial thickness burn?
   A. 5–10%
   B. 10–15%
   C. 10–20%
   D. 15–25%

15. The EMT must break blisters caused by a nonthermal burn to relieve the pain that the patient is experiencing.
   A. True
   B. False

16. How does shock affect the blood pressure and pulse?
   A. Increased pulse rate (early sign) and decreased blood pressure (late sign)
   B. Decreased pulse rate and increased blood pressure
   C. Strong, regular pulse and elevated blood pressure
   D. Irregular (early sign) pulse and increased (late sign) blood pressure

17. In addition to treating for shock, what special concern does an amputation injury present?
   A. Concern for airway compromise
   B. Concern for high blood pressure
   C. Concern for reattachment
   D. Concern for infection

18. Four mechanisms of injuries to the spine are compression, excessive flexion, extension and rotation, and lateral bending. A fifth mechanism is distraction, which is best described as:
A. Pulling apart of the spine.
B. Not being aware.
C. The removal of material used in a hanging.
D. All of the above

19. One of the signs and symptoms of a skull injury is bruising (discoloration) behind the ears. The ears are referred to as:
A. Proboscis.
B. Mastoid process.
C. Malkonian process.
D. Hearing organic process.

20. Which of the following is a precaution you should take in regard to cervical immobilization devices?
A. Wax all short and long boards for easier handling.
B. Maintain manual immobilization until the head is secured to the board.
C. Maintain manual immobilization throughout the entire transport after the long board is secure.
D. Both A and B are correct.

21. There are 17 general procedures for applying the long backboard. The first and last (1 and 17) are:
A. 1. Assess the cervical area; 17. Immobilize the head to the board.
B. 1. Assess pulses, motor, and sensory functions; 17. Tighten all straps.
C. 1. Start manual in-line immobilization; 17. Reassess pulses, motor, and sensation, and record.
D. 1. One EMT must maintain in-line immobilization; 17. Transport to the ER.

22. An EMT must remove a football or motorcycle helmet on all calls to provide proper care to the patient.
A. True
B. False

23. Rapid extrication is based on time and the patient, not on the EMT's preference.
A. True
B. False

24. When a patient is wearing a helmet, special needs for that patient include the fit of the helmet, the patient's movement within the helmet, the EMT's ability to gain access to the ABCs, and:
A. The patient's airway and breathing.
B. Check medic alert bracelet or chain.
C. Check for medical information card in patient's wallet or information from bystanders.
D. The patient's vital signs.

25. When an EMT provides emergency medical care for a head injury, body substance isolation is listed as the first procedure. Which of the following is not listed as the second, third, or fourth procedures?
A. Maintain the airway.
B. Complete the detailed physical exam en route, after completing the initial assessment with spine immobilization.
C. Control bleeding.
D. Immobilize the spine for suspected head injury.

Section

Infants and Children

Voices of Experience

Death in the Emergency Department

A 10-year-old boy named Joe was bicycling to school in a northern New England city along with another student and their teacher. He fell a little behind the others, and a city garbage truck made a left turn and struck him. He was brought to the hospital by the EMS crew; Joe was critically injured.

When Joe's parents arrived at the hospital, they were placed in a quiet room away from the resuscitation. Although they were Jewish, the hospital chaplain, who was not, was there with them.

The emergency department staff worked on Joe for hours. Finally, the attending pediatrics physician ordered the staff to stop the resuscitation. With great reluctance, they halted their resuscitation efforts.

Throughout the ordeal, the staff continuously relayed information to the parents and obtained medical information from them. The chaplain did not leave their side the entire time. After Joe was pronounced dead, his mother insisted on seeing him. The team advised against it, but she insisted. The team cleaned up the room and the boy, and then let the mother into the room, where she held him for a long time. During the day, those of us on the hospital medical staff who were parents found our way to the telephones and called home just to make sure everyone in our families was all right.

On the first anniversary of this tragedy, Joe's parents returned to the emergency department to talk to the staff and the chaplain. Their marriage had not survived the loss of their son.

In an ordeal such as this, it is important to provide emotional support to the family. The family should feel that they are part of the proceedings, and they should be allowed to spend as much time as they want with their child in order to cope with their loss. As care providers, we also feel a great sense of loss when we lose a child. Even if we know that we've put in our best possible effort, it is always difficult to lose a child.

All of us who were involved in the resuscitation efforts will carry Joe in our thoughts forever. During my years as an emergency department psychiatrist, I continued to talk about and relive that day. As hard as it is to remember such a painful memory, it helps, and the recollection becomes a little less painful with each telling. You will have patients like Joe who will live on in your heart. They will also live on in your mind and your hands and in your skills. Each EMT will deal with his or her "Joes" in different ways. For some of you, the pain may be too great, and you will pursue other paths. For others, the pain will help inspire you to be better at what you do and give you the strength to deal with the next injured child.

Stephen Michael Soreff, MD
President, Education Initiatives
Nottingham, New Hampshire
Faculty, Metropolitan College, Boston University
Boston, Massachusetts

SECTION

6 Infants and Children

"Never, never, never, never give up."
— Winston Churchill

Section 6 covers the following subject:

Lesson 6–1 Infants and Children

Presents information concerning the developmental and anatomical differences in infants and children, discusses common medical and trauma situations, and also covers infants and children dependent on special technology. Dealing with an ill or injured infant or child patient has always been a challenge for EMS providers.

Small airways throughout the respiratory system are easily blocked by:
Secretions and swelling of the airway.

The tongue is large relative to the small mandible and can block the airway in:
An unconscious child or infant.

Describe how the positioning of the airway is different in infants and children than in adults.
Do not hyperextend the neck.

Explain why suctioning a secretion-filled nasopharynx can improve breathing in an infant.
Because infants are obligate nose breathers.

Describe two ways children with dyspnea can compensate for short periods of time.
By increasing their breathing rate; by increasing the effort of breathing.

Respiratory compensation is followed rapidly by decompensation due to what two factors?
Rapid respiratory muscle fatigue; general fatigue of the infant.

Describe the technique used to open the airway in an infant or child.
Head tilt–chin lift; do not hyperextend.

What technique is used in opening a pediatric airway with suspected spinal injury?
Jaw thrust.

Describe the procedure for clearing a severe airway obstruction in infants less than 1 year old.
Back slaps/chest thrust and visual foreign body removal.

Describe the procedure for clearing a severe airway obstruction in responsive children older than 1 year.
Abdominal thrusts and visual foreign body removal.

What is the purpose of examining a child with the trunk-to-head approach?
To build confidence and to help relieve anxiety in the ill or injured child.

True or false: Toddlers will easily tolerate wearing an oxygen mask.

False.

Describe the three steps of inserting an oropharyngeal airway using a tongue blade.

Insert tongue blade to the base of the tongue; push down against the tongue while lifting upward; insert oropharyngeal airway directly without rotation.

Describe the proper technique for using a bag-mask.

Squeeze bag slowly and evenly enough to make sure the chest rises adequately.

What is the proper ventilatory rate for children and infants?

20 breaths per minute.

How can the EMT assess if he or she is properly using the bag-mask device?

Breath sounds should be heard in the lungs bilaterally.

General impression of a well as opposed to a sick child can be obtained from inspection of the overall appearance. List three signs that the EMT can use to obtain a generalized impression of the patient.

Mental status; effort of breathing; skin color.

When and where should the EMT begin assessing the pediatric patient?

As he or she enters; from across the room.

What factors can be assessed from across the room?

Mechanism of injury; condition of surroundings; general impression.

Seven items can be assessed to determine respiratory status in a pediatric patient. What are they?

Note chest expansion/symmetry; effort of breathing; nasal flaring; retractions; grunting; respiratory rate/quality; stridor, crowing, or noisy respirations.

Describe the proper technique for assessing circulation in infants and children.

Examine the brachial or femoral pulse, peripheral pulses, capillary refill, and skin color and temperature; blood pressure should be assessed in children older than age 3.

Stridor, crowing, or noisy respirations are all signs of what type of respiratory distress?

Partial airway obstruction.

True or false: Children with a partial airway obstruction should be removed from their parents and placed in a supine position.

False. Children should be assisted to the sitting position and can remain with the parents.

Describe how an EMT can deliver oxygen to a child who will not tolerate a nasal cannula or facemask.

Hold tubing 2 inches from the child's face or place the tubing into the bottom of a paper cup and hold it near the child's face.

An EMT must be able to recognize the difference between an upper airway obstruction and lower airway disease. Describe the different sounds an EMT must recognize in a child in respiratory distress.

Upper airway obstruction: stridor upon inspiration; lower airway disease: wheezing and breathing effort upon exhalation as well as rapid breathing (tachypnea) without stridor.

Severe airway obstruction is an extreme emergency. An EMT must be able to recognize the signs of severe airway obstruction. What are they?

No crying noise; no coughing noise; inability to speak; cyanosis.

A pediatric patient is said to be in respiratory arrest if his or her respiratory rate is less than _____ per minute.

10

When should oxygen be given to a child in respiratory distress?

Always.

The proper treatment for a child in respiratory distress and altered mental status will include the following:

Provide oxygen and assist ventilation with a bag-mask.

True or false: It is proper treatment to provide oxygen and assist ventilation in a child with the presence of cyanosis and poor muscle tone.

True.

True or false: Seizures in children who have chronic seizures are rarely life threatening.

True.

There are several medical causes of seizures in children. What are they?

Fever; infection; poisons; hypoglycemia; trauma; hypoxia.

Proper emergency medical care for the child having a seizure shall include:

Ensure airway position and patency; have suction ready; and protect from injury.

An EMT should provide__________ for the child having a seizure.

Oxygen.

An EMT may encounter a child with altered mental status. It is important for the EMT to recognize potential conditions that might cause this. List four.

Hypoglycemia; poisoning; post seizure; infection; head trauma; hypoxia.

Proper emergency medical care for the child with altered mental status shall include the following:

Ensure patency of the airway; provide oxygen; be prepared to suction and/or ventilate and transport.

A common environmental cause of infant and child ambulance calls is what?

Poisonings.

If at all possible, the EMT should do what with the container of a suspected poison?

Take it to the hospital with the patient for analysis.

Emergency treatment for a suspected poisoning should include what?

Contact medical control; consider need to administer activated charcoal; transport.

What is the first priority in treating an unresponsive child with a suspected poisoning?

Ensure patency of the airway and adequate ventilation.

A potentially life-threatening cause of fever in children is what?

Meningitis.

Identify some of the nontraumatic causes of shock in children.

Diarrhea; dehydration; vomiting; anemia; infection.

Describe the signs and symptoms of a child in shock.

Rapid respiratory rate; pale, cool, clammy skin; weak or absent peripheral pulses; delayed capillary refill.

Explain the rationale for rapid transport of an infant or child in shock.

Rapid transport of the patient ensures that definitive medical care can be given to the patient in shock within the "golden hour."

When should the secondary exam be performed on the infant or child in shock?

En route to the hospital.

Describe the proper emergency care that should be given to a child in shock.

Ensure open and patent airway; provide high-flow oxygen with a nonrebreathing mask; manage bleeding; keep the patient warm; provide rapid transport.

After the child is removed from the water, what is the top priority in near-drowning cases?

Artificial ventilation.

What is a frequent injury in diving accidents?

Cervical spine injury.

What is secondary drowning syndrome?

Delayed deterioration of the patient after breathing and pulse have returned; may occur from minutes to hours after the event.

At what age are most children at risk from dying of SIDS?

In the first year of life.

When are SIDS babies commonly discovered?

In the early morning.

What does SIDS stand for?

Sudden infant death syndrome.

How should the EMT deal with parents who have just suffered a SIDS death?

Avoid any comments that might suggest blame to the parents.

What is the leading cause of death in children and infants?

Trauma.

Of all the injuries suffered by children, which type is the most common?

Blunt injury.

The patterns of injuries as a result of motor vehicle accidents tend to differ among adults and children. Describe the type of injuries an EMT should expect in an *unrestrained* child.

Head and neck injuries.

Describe the type of injuries an EMT should expect in a *restrained* child.

Abdominal and lower spine injuries.

An ambulance crew is dispatched to a "child hit by a car." Upon arrival, the EMT learns the child was riding his bicycle when the automobile struck him. What three specific injuries should the EMT assess for?

Head injury; spinal injury; abdominal injury.

An ambulance crew is dispatched to a "child hit by a car." Upon arrival, the EMT learns the child was crossing the street when a car hit her. What specific injuries should the EMT assess for?

Abdominal injuries with internal bleeding; possible femur fracture; head injuries.

What is the single most important maneuver to ensure an open airway in a child with a suspected head and neck injury?

The modified jaw thrust.

What major complication should the EMT be assessing for secondary to head injuries in children?

Respiratory arrest.

The most common cause of hypoxia in the unconscious head injury patient is?

The tongue obstructing the airway.

What procedure should the EMT use to open an airway in an unconscious child?

Jaw thrust.

Why should the EMT avoid letting air into the stomach when ventilating a patient?

Air in the stomach can distend the abdomen and interfere with artificial ventilation efforts.

An EMT is transporting an injured child to the hospital. The child's condition continues to deteriorate but there are no external signs of injury. What major trauma should the EMT suspect?

Abdominal injuries with internal bleeding.

Describe the indications for use of the MAST pants.

Trauma with severe signs of shock and pelvic instability.

True or false: It is accepted practice when a child does not fit properly into the MAST pants to place them in just one leg.

False.

Proper treatment of burns by the EMT should include the following:

Cover with a sterile dressing or sheet.

What is the definition of child abuse?

Improper or excessive action so as to injure or cause harm.

What is the definition of child neglect?

Giving insufficient attention or respect to a child who has claim to that attention or respect.

What are the two common forms of child abuse that the EMT is likely to encounter?

Physical abuse; neglect.

The EMT needs to be able to recognize what could potentially be a case of child abuse and/or neglect. Describe common signs and symptoms of child abuse.

Multiple bruises in various stages of healing; injury inconsistent with the mechanism described; fresh burns; conflicting stories from the caregivers.

You and your partner respond to an address to which they have been four times within the last 8 weeks. Each time the victim is a child who lives in the home. On this response, you encounter a child who is unconscious and unresponsive. Physical exam reveals that the child has multiple bruises on his head and torso in various stages of healing. The parents tell you their son slipped off the couch about an hour ago and has not moved since. After treating the child and transporting him to the hospital, what other action must you take?

Suspect a case of child abuse and report it to the proper authority.

Of all the injuries sustained by abused children, CNS injuries are the most lethal. Describe what occurs to the child that suffers from shaken baby syndrome.

The child is violently shaken, causing injury to the cerebellum and medulla.

When an EMT suspects that a child may be a victim of child abuse, how should the EMT go about reporting the incident?

Do not accuse in the field; accusations and confrontation delay transport to the hospital. Bring objective information to the receiving hospital and relay it to the medical staff.

Today the EMT is very likely to encounter children with special needs. Often these children will be at home and technologically dependent. Identify some of the medical conditions of these special children.

The EMT may encounter premature babies with lung or cardiac disease; children with neurologic disease; children with chronic disease or altered function from birth.

What complications can arise from a tracheostomy tube?

Displacement; obstruction; bleeding; infection; air leakage.

What is the emergency medical care for a child with complications associated with the tracheostomy tube?

Maintain an open airway; ventilate and suction as needed; maintain a position of comfort; transport.

Home ventilators are more common than ever. Who can the EMT use as a resource when dealing with a child who is using one at home?

The parents are excellent resources for the EMT regarding the operation of the home ventilator.

Describe the possible complications an EMT may encounter in a patient with an IV site.

Site may be cracked, infected, actively bleeding, or may have clotted off entirely.

What is the proper treatment for a child with a nonfunctioning IV site?

If bleeding, apply pressure, and transport immediately.

What is a gastrostomy tube? What is its function?

A tube placed directly into the stomach; for feeding (these patients usually cannot be fed by mouth).

What is a VP (ventriculoperitoneal) shunt used for in a child's head?

To drain excess CSF from the brain; runs from the brain to the abdomen.

An ill or injured child can create incredible stress on parents as well as the EMT. Explain why striving for calm, supportive interaction with the family will result in improved ability to deal with the child.

Calm parents = calm child; agitated parents = agitated child.

How long should the EMT allow the parents to remain part of the child's care?

Parents should remain part of the child's care unless the child is unaware or medical condition requires separation.

Parents can be of great value in treating an injured or ill child. Describe how the EMT might utilize the parents in a positive manner.

Parents should be instructed to calm the child; they can maintain the child in a position of comfort; they can hold the oxygen device.

What can an EMT do to reduce his or her own stress when dealing with ill or injured children?

Advance preparation is important; practice with equipment and examining children to become more comfortable.

Identify a main area of stress an EMT may encounter when dealing with sick children.

An EMT may add additional stress by identifying the patients with their own children.

Infants and Children

1. There are two ways children with dyspnea can compensate for short periods of time. One is increasing their breathing rate, the other is:
A. Stop crying and breathe shallow breaths.
B. Breathing through their nose and closing their mouth.
C. Sitting themselves up, instead of lying down.
D. Increasing the effort to breathe.

2. In infants and children, respiratory compensation is followed rapidly by decompensation due to general fatigue of an infant and:
A. Rapid respiratory muscle fatigue.
B. General fatigue in older children.
C. Swelling of the nasal passages due to flaring.
D. Swelling of the trachea.

3. The procedure for clearing a severe airway obstruction in responsive children older than 1 year is:
A. Chest thrusts and back slaps.
B. Back slaps and finger sweeps.
C. Visual foreign-body removal and abdominal thrusts.
D. Visual foreign-body removal and back slaps.

4. Proper bag-mask use for infants and children requires the EMT to:
A. Squeeze the bag slowly and evenly so the chest rises adequately.
B. Squeeze the bag quickly so the chest rises adequately.
C. Squeeze the bag only halfway so the chest rises adequately.
D. Squeeze the bag quickly to cause the chest and stomach to rise and fall.

5. General impression of a child who is well as opposed to one who is sick can be obtained from inspection of overall appearance. Which of the following is not one of the three signs that the EMT can use to obtain a general impression of the child?
A. Skin color
B. Skin temperature
C. Mental status
D. Effort of breathing

6. When the EMT is delayed in reaching a patient, but is in the same room, which of the following factors cannot be assessed from across the room?
A. Mechanism of injury
B. Condition of the surroundings
C. General impression
D. Respirations and skin condition

7. You respond to a home, and the mother tells you that her 5-year-old child has a history of respiratory illness and cannot tolerate a nasal cannula or facemask. How can you provide oxygen to this child?
A. Hold oxygen tubing 2 inches from the child's face.
B. Place oxygen tubing into the bottom of a paper cup and hold it near the child's face.
C. Hold oxygen tubing no closer than 5–6 inches from the child's face.
D. Both A and B are correct.

8. For a pediatric patient, the EMT can use chest expansion, effort of breathing, nasal flaring, retractions, grunting, respiratory rate/quality, and stridor, crowing, or noisy respirations to assess the:
A. Mechanism of injury.
B. Severity of foreign body obstruction.
C. Respiratory status.
D. Respiratory rate and quality.

9. Children with a partial airway obstruction should be removed from their parents' laps or arms and immediately placed in a supine position.
   A. True
   B. False

10. Oxygen should be given to a child in respiratory distress:
   A. When vital signs are all above or below normal.
   B. In some instances.
   C. Always.
   D. Never when medical control unit is notified prior to use.

11. In children, fever, infection, poisons, hypoglycemia, trauma, and hypoxia indicate medical causes of:
   A. Seizures.
   B. Poisoning.
   C. Respiratory problems.
   D. Hyperactivity.

12. A pediatric patient would be in respiratory arrest if his or her respiratory rate is less than _______ per minute.
   A. 5
   B. 10
   C. 15
   D. 18

13. Proper emergency medical care for the child having a seizure includes ensuring airway position and patency, having suction ready, and:
   A. Providing oxygen at 15 liters per minute.
   B. Protecting the child from injury.
   C. Getting an AED to the scene.
   D. Preventing cyanosis.

14. Which of the following is not a nontraumatic cause of shock in children?
   A. Diarrhea and dehydration
   B. Vomiting
   C. Anemia and infection
   D. Partial airway obstruction

15. You just completed patient assessment of an infant or child and you treated any life-threatening problems. At what point do you complete the secondary survey?
   A. Immediately upon completion of the preliminary survey
   B. Continue ABCs; there is no secondary survey for the infant or child
   C. En route to the hospital
   D. Just before leaving the scene

16. You just removed a child from the water. What is the top priority in near-drowning cases?
   A. Get water out of the lungs.
   B. Begin artificial ventilation.
   C. Begin CPR.
   D. Attach AED pads to the child.

17. You are on duty early in the morning and returning from the hospital after a call. You are dispatched to a residence for an infant less than 1 year old, who is not breathing. You are advised that the infant is in a crib on the second floor of the residence, and the police on the scene tell you that there are no signs of trauma. What might you be responding to?
   A. Homicide
   B. Sudden death
   C. Sudden infant death syndrome
   D. Foreign-body airway obstruction

18. Of all the injuries suffered by children, which type is the most common?
A. Water-related injuries
B. Trauma
C. Blunt trauma
D. Choking or food-related trauma

19. You respond to a "child hit by a motor vehicle." The child was riding his bicycle when he was hit by the motor vehicle. Which of the following is not one of the three specific injuries you assess for?
A. Head injury
B. Spinal injury
C. Abdominal injury
D. Possible femur fracture

20. The EMT responding to a child with a suspected head and neck injury must ensure an open airway by:
A. Head tilt–chin lift.
B. Cervical collar and oropharyngeal airway.
C. Modified jaw thrust.
D. Tongue–jaw lift.

21. Blunt trauma is most common in children and it is one of the reasons why children may require CPR.
A. True
B. False

22. Initially, ensure a patent airway, provide supplemental high-flow oxygen, and be prepared to suction and assist with ventilation if necessary. Transport to the hospital as soon as possible for a patient with:
A. Confusion.
B. An altered mental status that is not related to hypoglycemia.
C. An altered mental status that is directly related to hypoglycemia.
D. An altered mental status that is directly related to dyspnea.

23. For an emergency involving a poisonous substance, bring all containers, bottles, etc., only if:
A. All containers are sterilized.
B. All containers are marked with permanent marker.
C. It can be done safely.
D. They are known substances in their original containers.

24. Oral glucose assists with the:
A. Proper function of the brain.
B. Proper pancreas function.
C. Proper liver function.
D. Proper blood circulation.

25. Regarding an allergic reaction, which of the following would not be a sign or symptom when assessing the patient's respiratory system?
A. Coughing
B. Noisy respirations
C. Wheezing or stridor
D. Itchy, watery eyes

Section

Operations

Voices of Experience

Is the Scene Really Safe?

It was Thursday, November 29, at approximately 2100. I was serving as the battalion EMS officer for the shift. I was dispatched on a run to a possible stroke victim in the city of Columbus with a township medic and engine, because all of my battalion medics were tied up. En route, the FAO called and said the call was now a cardiac arrest and CPR was in progress. A minute or two later, the township medic came on the radio and said they were on scene with the engine and that they had a full arrest. I arrived about 5 minutes later. The front door was open and the engine officer was busy running in and out of the residence retrieving various supplies, apparently for the medics inside. I got out of my unit and walked toward the house. As I reached the front door, I noticed a strong odor or exhaust coming from the house. As I stepped through the door, the odor was so strong I could actually taste it on my lips. I walked into the living room, down a hallway to the left, and into a bedroom on the right side of the hallway to find five fire department personnel working on this apparent arrest patient.

I asked if anyone noticed the strong odor, and the response from two of the medics was, "We are a little dizzy." To which I responded, "That would be the *CARBON MONOXIDE*!" I told them to open some windows and get the victim and themselves out of the building while I called in a CO run. The ladder, rescue, engine, and battalion chief arrived on the scene and did their thing. In the meantime, the victim's brother walked out of the house into the front yard and began to vomit and complain of dizziness.

I had another medic evaluate him and transport him to the emergency department. The five fire department personnel and the arrest victim went to the hospital as well. The arrest victim stayed arrested with a carboxyhemoglobin of 64. The medics were evaluated and found to have a count of between 11 and 18. The emergency department physician mentioned that it should never get over 9 to 9.5. The rescue crew got readings of from 150 parts per million outside the front door to 934 ppm inside the house at least 20 minutes later. Standard procedure requires evacuation of a residence at 35 ppm.

We had a critique the following workday with all who were on the scene. At that meeting, the township crew said they would not do anything differently if they had to do it again. I mentioned that perhaps they might recognize the odor of exhaust or the dizziness the next time. At least after this time, there might be a next time.

Scene safety is the number one priority that we are taught in our EMS training. Why do we forget that in situations where we need to remember it the most?

Lieutenant Dan McConnell
EMS Coordinator
Columbus Ohio Division of Fire
Columbus, Ohio

SECTION

7 Operations

"In medicine sins of commission are mortal, sins of omission are venial."
— Theodore Tronchin

Section 7 covers the following subjects:

What is the minimum staffing level in the ambulance patient compartment?

One EMT.

What is the preferred staffing level in the ambulance patient compartment?

Two EMTs.

Which mechanical items are you required to inspect on the ambulance daily?

Fuel; oil; engine-cooling system; brakes; wheels and tires; door closing and latching system; air conditioning/heating system; ventilation system.

Which electrical systems are you required to inspect on the ambulance daily?

Horn; wipers; headlights; stoplights; turn signals; emergency warning lights; siren; communications system.

In addition to mechanical and electrical systems, what else must you check on the ambulance daily?

Equipment should be checked and maintained, restocked and repaired; batteries for defibrillator charged; suction working; oxygen filled.

What three elements comprise a good dispatch system?

Central access; 24-hour availability; trained personnel.

What dispatch information must be obtained from a caller?

The nature of the call; name; location and callback number of the caller; location of the patient; number of patients and severity; other special problems.

Mike and Cathy mount up for a call. What is the first thing Mike checks about Cathy and himself?

That they are properly buckled up.

When is the first call made to dispatch from the ambulance?

When it is en route.

What must the dispatcher inform the responding unit?

The nature and location of the call.

Must all ambulance drivers attend an approved driving course?

Mandated only in some states.

What are the characteristics of a good ambulance operator?

Physically fit; mentally fit; able to perform under stress; has a positive attitude about his or her abilities; is tolerant of other drivers.

It is three o'clock in the morning. Lieutenant Lane prepares to depart the hospital to take Mrs. Blum back to her home from the emergency department 2 hours after he brought her in for a headache. Who must wear seatbelts for the ride home?

The driver and all passengers.

What should Paramedic Tracy learn before driving the mobile intensive care unit for the first time?

Become familiar with its special characteristics.

It is Sunday, January 17. Medic 310 hasn't had a call all day, and Mike, who is driving today, has been sacked out on the couch since checking out the vehicle this morning. It is now 1845 hours and he is awakened for a call. As the bay doors open, what is he looking for?

In addition to traffic, any changes in weather and road conditions that might affect his driving.

How would you respond to the following: EMTs get extra protection from using lights and sirens because it is easier for other drivers to see and hear them.

Extreme caution must be exercised in the use of lights and sirens; their use considerably increases the hazards to the crew.

"Medic 86, respond to Villa Chameleon. Possible blocked Foley catheter." John is driving. What must he consider?

He must select an appropriate route for current conditions.

True or false: When responding to an emergency, one need not maintain the normal following distance.

False. One must always maintain a safe following distance.

True or false: When responding to an emergency, the driver is responsible only to his or her crew.

False. Drive with due regard for the safety of all others.

Right out of EMT school, Tim uses lights and sirens any time he gets a call from dispatch, arguing any call might ultimately be an emergency. Comment.

This is unsafe practice; every EMT must know when it is appropriate to use lights and sirens.

What are the most visible warning lights on an emergency vehicle?

Its headlights.

Prior to arriving on scene, what additional considerations should the squad leader have considered?

Obtaining additional information from dispatch; assigning personnel to specific duties; assessing specific equipment needs; how to position the vehicle when on scene.

What specific considerations should the squad driver take before positioning the vehicle?

Uphill from leaking hazards; upwind from airborne hazards; 100 feet from wreckage; avoid parking in a location that will hamper exit from the scene.

What should the driver do before exiting the vehicle?

Set the parking brake; utilize warning lights; shut off the headlights unless there is a need to illuminate the scene.

What are some of the local laws, regulations, and ordinances relating to the operation of emergency vehicles that you should review?

Vehicle parking or standing; procedures at red lights, stop signs, and intersections; regulations regarding speed limits; direction of flow or specified turns; emergency or disaster routes; use of audible warning devices; use of visual warning devices; school bus safety.

What is the most dangerous mode of emergency response?

Escorts and multiple vehicle response.

When might you use a multiple vehicle response?

Only if you are unfamiliar with the location of the patient or the receiving facility.

What special precautions should you take if you must utilize a multiple vehicle response?

No vehicle should use lights or siren; provide a safe following distance; recognize the hazards of multiple vehicle responses.

What is the most common type of crash?

Intersection crashes.

What are some of the causes of intersection crashes?

Motorists arriving at an intersection as the light changes and not stopping; multiple emergency vehicles following closely and waiting motorists do not expect more than one; vision obstructed by other vehicles.

What is the first thing you need to do when arriving on scene and assuring scene safety?

Notify dispatch.

What are the elements of scene size-up?

Body substance isolation; scene safety; mechanism of injury/illness; total number of patients; the need for additional assistance.

When should you consider using body substance isolation?

Prior to patient contact.

What types of body substance isolation should you have available?

Gloves; gowns; eyewear.

What are the three primary aspects of scene safety you should consider?

Is the emergency vehicle parked in a safe location? Is it safe to approach the patient? Does the victim require immediate movement because of hazards?

You and your partner are first on scene of an RV that has rolled on the Ohio Turnpike. There are victims lying all over the median. What are the first three things you should do as incident commander?

Determine the number of patients; obtain additional help; begin triage.

What adjectives would you use to describe appropriate actions at an MCI?

Organized; rapid; efficient.

To what end should your actions at an MCI be directed?

Organized, rapid, efficient transport.

What actions should you take when preparing a patient to be transferred to the ambulance?

Complete all critical interventions; check dressings and splints; cover patient; secure to stretcher.

Jonathan has just taken off in the helicopter to transfer a patient from Macon to Atlanta. What six things should he aim to accomplish during the transport?

Notify dispatch; continue an ongoing assessment; obtain additional vital sign measurements; notify the receiving facility; reassure the patient; complete the prehospital care report.

What is the first thing the EMT should do on arrival at the receiving facility?

Notify dispatch.

What two types of reports should be given to the receiving facility?

Verbal report at bedside; written report completed and left prior to returning to service.

What tasks do you need to accomplish to prepare for the next call?

Clean and disinfect the ambulance and equipment as needed; restock the disposable supplies; refuel the unit; file reports; notify dispatch.

What considerations should you take when calling for air transport?

Proper utilization of aircraft; landing zones; safety.

Mike and Dave roll up on the scene of an MVA at US 57 and Bell. Elyria Fire is on scene, and they are preparing to extricate the patient. As senior EMT on scene, what is Mike's role?

To administer necessary care to the patient before extrication and ensure that the patient is removed in a way to minimize further injury. Patient care precedes extrication unless delayed movement would endanger the life of the patient or rescuer. He must also ensure that he and his crew work well with the other emergency personnel on scene.

In working with the fire department personnel, what must Mike ensure?

That the nonrescue EMS providers work together with the providers of rescue, but not allow their activities to interfere with patient care.

You roll up in Medic 86 at a car-versus-pole scene at 9th Avenue, and Middle Elyria Fire radios that the rescue vehicle is at another crash. An engine has been dispatched but has been delayed at a train crossing and will be there in 15 minutes. What principles must you keep in mind as both rescuer and EMT?

Establish a chain of command to ensure patient care priorities; administer necessary care to the patient before extrication; ensure that the patient is removed in a way to minimize further injury. Patient care precedes extrication unless delayed movement would endanger the life of the patient or the rescuer.

What is the number one priority for all EMS personnel?

Safety.

What kind of protective clothing should be utilized?

Whatever is appropriate for a given situation.

Following the safety of the EMS responders, what is the next priority?

The safety of the patient.

During extrication, what should you tell the patient?

The patient should be informed of the unique aspects of extrication.

From what should you protect the patient during extrication?

Broken glass; sharp metal; and other hazards, including the environment.

What can you do to gain vehicle access that does not require equipment?

Try opening each door; roll down windows; have patient unlock doors.

What are examples of courses that can be taken for learning the techniques of complex access?

Trench; high angle; vehicle extrication; confined spaces; SCBA; swift water; SWAT; etc.

What is the sequence of events in treating the victim of an MVA once access is gained?

Maintain cervical spine stabilization; complete the initial assessment; provide critical interventions; immobilize the spine securely (unless rapid extrication is necessary); move the patient—not the immobilization devices; use sufficient personnel; choose the path of least resistance; continue to protect the patient from hazards.

For whose safety is the EMT concerned at a hazardous materials incident?

His or her own; the crew's; the patient's; the public's.

When approaching a scene, how can you find out if hazardous materials are involved?

By knowing the occupancy of the building or vehicle; looking at the size and shape of containers; looking for placards and shipping papers; using your senses.

What are some general procedures you should use when approaching the scene of a possible hazardous materials incident?

Park upwind/uphill from the incident at a safe distance; keep unnecessary people away from the area; isolate the area by keeping people out; do not enter without proper equipment; avoid contact with the material; remove patients to a safe zone if no risk to EMT. Do not enter the hazardous materials area unless trained as a hazardous materials tech with proper training and SCBA.

In addition to the possible hazardous materials and equipment, what is another vital consideration when approaching a hazardous materials incident?

Environmental conditions and hazards.

What resources should you be aware of when dealing with a hazardous materials incident?

Local hazardous materials response teams; CHEMTREC 800-424-9300; *Hazardous Materials, The Emergency Response Handbook,* published by the U.S. Department of Transportation.

Where can you find the National Fire Protection Association (NFPA) Hazardous Materials requirements for EMS providers?

NFPA 479 and OSHA 1910.120

What is an incident management system?

A system developed to assist with the control, direction, and coordination of emergency response resources; provides an orderly means of communication and information for decision making; makes interactions with other agencies easier because of the single coordination.

After an incident manager is determined, what seven EMS sectors are established as needed?

Extrication; treatment; transportation; staging; supply; triage; mobile command center.

On his way home from the station, Captain Haskell heard a loud explosion and saw a fireball in the distance. When he arrived on the scene of the chemical plant explosion, to whom did he report?

The sector officer for specific duties.

Once assigned his task by the sector officer, what should he do?

Complete the task, and then report back to the sector officer.

What is the definition of a multiple-casualty situation?

An event that places great demand on resources, be it equipment or personnel.

What is basic triage?

Sorting multiple casualties into priorities for emergency care or transportation to definitive care.

What are the three triage levels?

Highest priority; second priority; lowest priority.

What complaints are included in the highest priority?

Airway and breathing difficulties; uncontrolled or severe bleeding; decreased mental status; patients with severe medical problems; shock (hypoperfusion); severe burns.

What complaints are included in the second priority?

Burns without airway problems; major or multiple bone or joint injuries; back injuries with or without spinal cord damage.

What complaints are included in the lowest priority?

Minor painful, swollen, deformed extremities; minor soft-tissue injuries; death.

Who should become the triage officer?

The most knowledgeable EMS provider arriving on scene first.

What are the responsibilities of the triage officer?

Request additional help; perform initial assessment on all patients; assign available equipment and personnel to priority one patients; prioritize patient transport; remain at the scene to assign and coordinate personnel, supplies, and vehicles.

What factors must be taken into account when making patient transport decisions?

Prioritization; destination facilities; transportation resources.

Operations

1. Central access, __________, and __________ are the three elements that comprise a good dispatch system.
   A. 24-hour availability; experienced call takers
   B. 24-hour availability; trained EMTs
   C. 24-hour availability; trained personnel
   D. 24/7 availability; advanced training

2. Getting ready to respond to a call with the ambulance is often referred to as "mounting up." What must the two responding EMTs check about each other?
   A. That they are properly buckled up
   B. That their lights are turned on
   C. That their ambulance radio is turned on
   D. That they both know the exact location of the call

3. When working a 24-hour shift, the ambulance driver is awakened 12 hours after a mid-afternoon call. As the ambulance driver prepares to exit the building, the driver should make:
   A. Sure all lights work.
   B. A safe entrance to traffic and pay attention to weather and road conditions.
   C. Sure that the brakes are tested.
   D. Sure that the rear door is secure.

4. The most visible warning lights on an emergency vehicle, including an ambulance, are the:
   A. Visi-bar lights.
   B. Halogen back-up lights.
   C. Red strobe lights.
   D. Headlights.

5. Extreme caution must be exercised with the use of lights and sirens, because their use can cause:
   A. Other vehicles to drive erratically to move to the side of the road.
   B. The hazards to the crew to increase considerably.
   C. Temporary blindness to other vehicle operators.
   D. Sirens to be more difficult to hear with long-time users.

6. Before positioning the vehicle, the squad driver should consider several things, which include parking in a location that will not hamper exit from the scene.
   A. True
   B. False

7. Regarding Question 6, another consideration is to park downward from airborne hazards.
   A. True
   B. False

8. The most dangerous modes of emergency response include escort and:
   A. Areas of roadwork.
   B. Multiple vehicle response.
   C. Rush-hour traffic.
   D. Entering intersections.

9. The most common type of crash is:
   A. Rear-end crashes.
   B. Slippery road crashes.
   C. Crashes at intersections.
   D. Crashes at stop signs.

10. After the driver properly positions the ambulance, he or she should set the parking brake, utilize warning lights, and:
- **A.** Turn off the visi-bar lights.
- **B.** Shut off the headlights unless they are necessary to illuminate the scene.
- **C.** Turn off the wig-wag lights, but leave the headlights on.
- **D.** Be sure that the high beams are left on day or night.

11. Intersection crashes occur during multiple vehicle responses, because waiting motorists do not expect more than one emergency vehicle to enter the intersection.
- **A.** True
- **B.** False

12. What is the first thing you need to do when arriving on the scene and ensuring scene safety?
- **A.** Notify dispatch.
- **B.** Take body substance isolation precautions.
- **C.** Park downhill from toxic spills.
- **D.** Check your portable radio.

13. Three important types of body substance isolation for the EMT are:
- **A.** Protective goggles, gloves, and eyewear under goggles.
- **B.** Eyewear, heavy-duty rubber gloves, and an apron.
- **C.** Foot covers, latex gloves, and eyewear.
- **D.** Nonlatex gloves, gowns, and eyewear.

14. What must you do before beginning triage at a multiple injury scene?
- **A.** Determine the number and type of injuries.
- **B.** Call dispatch with an accurate request for additional help.
- **C.** Determine the number of patients and obtain additional help.
- **D.** Determine priorities and tag the patients by priority, and get wreckers and utilities en route.

15. You arrive at the receiving facility with your patient. What should the EMT vehicle operator do first?
- **A.** Position the vehicle and turn off the lights.
- **B.** Notify dispatch.
- **C.** Notify the receiving facility that you have arrived.
- **D.** Secure your vehicle.

16. If fire personnel have properly prepared a patient for extrication, and you and your partner are the senior EMTs on the scene, what should you do before the actual extrication?
- **A.** Assume the incident commander role.
- **B.** Notify the senior fire officer that you are in charge and will take over the extrication.
- **C.** Administer necessary care to the patient before extrication and ensure that the patient is extricated in a way that minimizes further injury.
- **D.** Ensure that others on the scene stay within their policies and procedures and do not interfere with extrication.

17. What is the number-one priority for all EMS personnel?
- **A.** Unity of command
- **B.** Unified command
- **C.** Body substance isolation precautions
- **D.** Safety

18. EMTs give two reports to the receiving facility: a ____________ report at bedside and a ____________ report completed and left prior to returning to service.
- **A.** Verbal; written
- **B.** Written; written
- **C.** Formal; hand-written
- **D.** Run-sheet; completed run-sheet

19. Several tasks need to be completed to prepare for the next call: clean and disinfect the ambulance and equipment as needed, restock the disposable supplies, refuel the unit, file reports, and:
A. Get patient information to other assisting departments.
B. Notify dispatch.
C. Phone in unusual incidents to agency lawyers.
D. Refresh, rehydrate, and change uniforms or coveralls.

20. Patient care precedes extrication unless:
A. Delayed movement would endanger the life of the patient.
B. Delayed movement would endanger the life of the patient and rescuer.
C. More patients are at the scene.
D. None of the above; patient care always precedes extrication.

21. One of the easiest ways to gain access to a patient in a vehicle that does not require equipment is to:
A. Try opening each door.
B. Have the patient unlock the doors.
C. Roll down the windows.
D. All of the above

22. At a hazardous materials incident, whose safety is the EMT concerned about?
A. His or her own
B. The patient
C. The public
D. All of the above

23. In addition to the possible hazardous materials and equipment, what other vital considerations must be taken on approaching the incident?
A. Wearing proper protection
B. Parking downwind from incident
C. Environmental conditions and hazards
D. Extrication requirements

24. During incident command (IMS), the EMT works within unity of command and is responsible directly to the.
A. Incident commander.
B. Sector officer.
C. Fire captain.
D. Hazardous materials incident commander.

25. Who should become triage officer?
A. The first EMT on the scene
B. Dual responsibility of the first two EMTs arriving on the scene in the first ambulance
C. The most knowledgeable EMT arriving on the scene
D. The first chief or deputy on the scene

Section

8 Advanced Airway (Elective)

Voices of Experience

The Breakfast Code

Jennifer and I were just starting our shift on Medic 208. It was about 6:50 A.M. and we were heading to the garage to check out the truck. Medic 202 got toned out to an unresponsive person, and we jumped the run so they could clock out on time, as their relief crew had not yet arrived.

I was driving and Jennifer and I were talking about the fact that Squad 301, our supervisor unit, had not been toned out on this run with us. As we approached the address, we were looking for the correct house number when Bill Cherry, another medic, stuck his head up through the divider between the cab and the back of the rig and said, "I think it is this house here on the right." Both Jennifer and I just about jumped out of our respective windows, because neither of us knew that he was in the back. He said, "I thought I would just tag along on this run in case it is a code." It did indeed turn out to be a code to remember.

We knocked on the front door and the patient's husband opened it and said, "She's in there. She's done it again!" We walked into the bedroom that was just off the kitchen and found the patient, a female around 65 years of age. She was lying on the bed with her feet hanging over the right side. She was pulseless and apneic, but still warm and pink.

We pulled the patient off the bed and into the kitchen, just a few feet away, where there was more room. We began CPR, got the monitor ready, and placed the paddles on the patient's chest. She was in V-fib. We shocked her three times. She was still in V-fib. While we continued CPR, got ready to intubate her, and prepared the IV tubing, I noticed her husband walking around the kitchen.

It was a big kitchen with an island that separated it from the dining area. There were two entrances to the kitchen; one at the head of the patient and one near her feet. Her husband would go out one door and in the other. At first I thought he was just nervous and trying to see what we were doing, but then I smelled fresh coffee brewing and realized that he had just made his morning coffee.

Bill was intubating, Jennifer was getting the IV started, and I was doing compressions, when I felt a tap on my shoulder. I looked back and the husband asked, "Would you like a cup of coffee?" We all looked at one another, and I said to the husband, "Actually I would, but I'm a little busy right now!" "She'll be OK," he said. "She's done this before."

We all looked at one another again and went on with what we were doing. I pushed some epinephrine and defibrillated again. Then I pushed some lidocaine and defibrillated again. CPR continued and we repeated the epi and lidocaine. After we shocked her the sixth and seventh time, she converted to a wide-complex rhythm with pulses.

We smelled something cooking and glanced up at the husband, who had been walking in and out of each of the doors. He was cooking a full breakfast of eggs, bacon, hash brown potatoes, and toast. He looked down at us with a spatula in one hand and a plate in the other and asked, "Do you guys want something to eat?" By this time, it was hard to keep from laughing out loud.

Just then, the patient went back into V-fib, and we shocked her again. This time she went into a sinus rhythm with pulses and a good blood pressure. We were trying to get her onto a long backboard so we could get her onto the cot more easily. Every time we tried to move her, she would throw a different rhythm at us that we would have to attend to. Once again the husband asked if we wanted anything like coffee or some bacon. He said, "I can put it between a couple pieces of toast and maybe add a fried egg?" We were very tempted, considering that we had been on the floor for more than half an hour working on the patient, but we graciously refused.

Finally, our patient was stable enough to load onto the cot. Bill and Jennifer lifted her onto the cot while I continued to bag. I tried to stand up, but found that I had been on my knees for so long that I couldn't. Jennifer and Bill both had to help me up! We loaded up into the ambulance and Bill drove us to the hospital. By the time we got to the emergency department, our patient was having some spontaneous breathing efforts and her blood pressure was good.

She was placed on a respirator in the emergency department and taken up to ICU after about half an hour. We finally got a look at her old chart and found out that she had indeed coded before. I went up to see her the next day, and she had already been extubated. Her family thanked me, and I told the patient that I was there to help her. She just smiled and didn't say anything. In hindsight, seeing how well everything turned out, I guess we could have at least gotten breakfast to go!

Dave Linton, EMT-P/Instructor
Bloomington Hospital Ambulance Service
Pellham Emergency Training, Inc.
Bloomington, Indiana

8 Advanced Airway (Elective)

"By medicine life may be prolonged, yet death will seize the doctor too."
— William Shakespeare

Section 8 covers the following subject:

Lesson 8–1 Advanced Airway

Instructs students on how to maintain an airway by means of orotracheal intubation. Included is a review of basic airway skills, nasogastric tube insertion for decompression of the stomach of an infant or child patient, and orotracheal intubation of adults, infants, and children. This lesson should be presented prior to the medical and trauma modules.

NOTE: The student should consult with his or her instructor or state EMS to determine if this module of instruction is included in the state curriculum.

Identify the following: a leaf-shaped structure that prevents food and liquid from entering the trachea during swallowing.

Epiglottis.

Identify the following: firm cartilage ring forming the lower portion of the larynx.

Cricoid cartilage.

Identify the following: two major branches of the trachea to the lungs.

Bronchi.

Describe the active process of inhalation.

Diaphragm and intercostal muscles contract, increasing the size of the thoracic cavity; ribs move upward and outward; air flows into lungs.

Describe the process of air exchange at the alveolar/capillary level.

Oxygen-rich air enters the alveoli during each inspiration; oxygen-poor blood in the capillaries passes into the alveoli; oxygen enters the capillaries as carbon dioxide enters the alveoli.

What is the normal breathing rate for an adult?

12–20 breaths/min.

What is the normal breathing rate for a child?

15–30 breaths/min.

What is the normal breathing rate for an infant?

25–30 breaths/min.

The EMT must be able to identify normal respiratory patterns in order to identify abnormal respiratory patterns. Describe the normal respiratory pattern.

Rate and rhythm is regular; breathing effort is unlabored without the use of accessory muscles; breath sounds are present bilaterally; chest expansion is adequate and equal.

List some signs of inadequate breathing.

Rate is outside normal limits for the age of the patient; breath sounds are diminished or absent; respiratory effort is increased with use of accessory muscles; tidal volume may be inadequate or shallow; nasal flaring may be present; the skin may be pale or cyanotic and cool and clammy; agonal breathing may be seen just before death.

Identify some of the differences between the adult and pediatric airway.

All structures in general are smaller and more easily obstructed in children; infants' and children's tongues are proportionally larger; the trachea is narrower and so easily obstructed by swelling; the trachea is softer and more flexible in infants and children.

What is the narrowest area in a child's airway?

The cricoid cartilage region.

What is the best method for opening an airway?

The head tilt–chin lift when there is no suspicion of neck injury.

What is the best method for opening an airway in a patient with a suspected neck injury?

The jaw-thrust maneuver.

What is the purpose of suctioning?

To remove blood, other liquids, foreign objects, and food particles from the airway.

When should a patient be suctioned?

Immediately when a gurgling sound is heard.

What are some complications of orotracheal suctioning?

Cardiac arrhythmia; hypoxia; coughing; mucosal damage; bronchospasm.

What is the function of a nasogastric tube?

To decompress the stomach and proximal bowel in cases of obstruction or trauma; for gastric lavage in the presence of upper GI ingestion or bleeding; for the administration of medications and nutrition.

What are the indications for using a nasogastric tube in the prehospital setting?

Inability to artificially ventilate the infant or child because of gastric distention.

What is the proper size of nasogastric tube that can be inserted in a newborn or an infant?

8.0 French.

What is the proper size of nasogastric tube that can be inserted in a toddler/preschooler?

10.0 French.

What is the proper size of nasogastric tube that can be inserted in a school-age child?

12 French.

What is the proper size of nasogastric tube that can be inserted in an adolescent?

14–16 French.

Describe the procedure for measuring the proper length of a nasogastric tube.

Measure tube from the tip of the nose, around the ear, to below the xiphoid process.

Occasionally the EMT may need to use a technique called the "Sellick maneuver." Describe this technique and the indications for its use.

A slight pressure placed on the cricoid cartilage when intubating a patient; purpose of the maneuver is to prevent passive regurgitation and aspiration during endotracheal intubation; may also aid in visualizing the vocal cords.

Where should an EMT look to find the cricoid cartilage?

The depression below the thyroid cartilage (Adam's apple) is palpated. This corresponds to the cricothyroid membrane.

What is the purpose of orotracheal intubation?

It is the most effective means of controlling a patient's airway; it is used in apneic patients; it provides good control of the airway and minimizes risk of aspiration; it allows for better oxygen delivery and deeper suctioning.

Describe the possible complications that may result from orotracheal intubation.

A slowing of the heart rate may occur as a result of stimulating the airway; it may cause damage to the lips, teeth, gums, and airway structures; it may cause vomiting, esophageal intubation, and right main-stem intubation.

What is the primary cause of extubation in infants and children?

Movement is the primary cause of extubation. Be sure to reassess breath sounds following every move; e.g., from the scene to the ambulance, from the ambulance to the receiving hospital.

Which style of laryngoscope blade is preferred in children?

The straight blade.

Describe the proper technique when using a straight blade for intubation.

Lift the epiglottis to allow visualization of the glottic opening and vocal cords; place the tube into the trachea.

Assorted sizes of endotracheal tubes are available for patient use. What size of ET tube should the EMT place in an adult male?

8.0–8.5 mm.

Assorted sizes of endotracheal tubes are available for patient use. What size of ET tube should the EMT place in an adult female?

7.0–8.0 mm.

Assorted sizes of endotracheal tubes are available for patient use. If limited equipment is available, what size of ET tube could be used in an emergency?

Emergency rule: 7.5 fits any adult.

Describe the components of an endotracheal tube and their functions.

15 mm adapter—allows for attachment of bag-mask; pilot balloon—verifies the cuff is inflated; cuff—holds approximately 10 cc of air to provide a seal.

What is the main difference between adult and child ET tubes, other than size?

Most infant and child endotracheal tubes are uncuffed.

What is the maximum age of a patient in which an uncuffed ET tube is used?

8 years old.

What is the "Murphy eye" of the ET tube?

The small hole on the left side across from the bevel that decreases the chances of obstruction.

What is the average distance from the teeth to the vocal cords?

15 cm.

What is the average distance from the teeth to the sternal notch?

20 cm.

What is the average distance from the teeth to the carina?

25 cm.

What device can be used to help shape the ET tube?

A stylet.

How far should a stylet be inserted into the ET tube?

The Murphy eye.

Identify several indications for intubation.

The inability to ventilate an apneic patient; patient with no gag reflex; the inability of the patient to protect his or her own airway, e.g., cardiac arrest, unresponsiveness.

Elena, working the Combat Zone, is responding to a report of "an unconscious person who fell from a rooftop." What should she keep in mind when intubating this patient?

If trauma is suspected, the patient must be intubated with the head and neck in a neutral position using in-line stabilization.

Antonio, senior medic on Squad 1, is preparing to intubate a cardiac arrest patient. What is the proper landmark he should use for insertion of a curved blade?

The curved blade should be inserted into the vallecula and lifted upward.

Sam, right out of the paramedic academy, is preparing to intubate an overdose patient. What is the proper landmark he should use for insertion of a straight blade?

The straight blade is used to directly lift the epiglottis, then insert the ET tube.

Ensuring proper placement of the ET tube is absolutely essential. Describe the technique that Captain Veronica, the senior EMT, should use after intubating a trauma patient to ensure proper tube placement.

Auscultate breath sounds; begin over the epigastrium; no sounds should be heard during artificial respiration. Listen over the left apex; compare with the right apex; breath sounds should be heard bilaterally.

Your lieutenant, Jacob, just intubated a patient. While assessing breath sounds for him, you notice they are slightly diminished on the left side of the chest. What does this finding indicate?

If breath sounds are diminished or absent on the left, most likely a right main-stem intubation has occurred.

What is the concern with placing an ET tube into the abdomen?

An unrecognized esophageal intubation is fatal.

There are several potential complications that may be encountered while intubating an adult patient; identify them.

Trauma to the lips and teeth, tongue, gums, other airway structures; prolonged attempts may lead to hypoxia; right main-stem intubation; esophageal intubation; self-extubation; vomiting.

Where is the narrowest part of a child's airway?

The cricoid cartilage.

Diana, an EMT student, is preparing her equipment to intubate an infant. What is the preferred laryngoscope blade to use in an infant?

A straight blade; provides greater displacement of the tongue.

Bob, 2 days before retirement after 20 years on the job, is restocking his airway bag after a critical run. When checking his supply of ET tubes, he sees several were lost at the scene. What size of ET tube should he have for newborns and small infants?

3.0–3.5.

Cheri, the flight medic on scene of a school bus rollover, is attempting to find the proper ET tube size for use in an unconscious child. What are two methods she can use to determine proper ET tube size?

Size of the little finger; the same size as the patient's nare.

Advanced Airway (Elective)

1. The leaf-shaped structure that prevents food and liquid from entering the trachea during swallowing is the:
   A. Ericoid.
   B. Epiglottis.
   C. Pharynx.
   D. Thoracic bronchi.

2. During inhalation, the diaphragm and intercostals muscles contract, and the ____________ increases in size.
   A. Thoracic cavity
   B. Bronchi
   C. Epiglottis
   D. Nasal pharynx

3. The patient has "major branches" called bronchi, located in the:
   A. Left ventricle.
   B. Right intercostal space.
   C. Lungs.
   D. Stomach.

4. The ____________ is located on the lower portion of the larynx.
   A. Trachea
   B. Epiglottis
   C. Pharynx
   D. Cricoid cartilage

5. EMTs must be able to identify normal respiratory patterns in order to identify abnormal respiratory patterns. Normal chest expansion is described as:
   A. Adequate and equal.
   B. Unlabored and using accessory muscles.
   C. Equal and regular.
   D. Bilateral and equal.

6. In relation to inadequate breathing, the EMT would see agonal breathing:
   A. And bilateral movement.
   B. Prior to breathing being restored to normal.
   C. Along with nasal flaring.
   D. Just before death.

7. The best method to open an airway is the head tilt–chin lift maneuver UNLESS:
   A. There is more than one rescuer.
   B. There is suspicion of a neck injury.
   C. The patient is an infant.
   D. A nonrebreathing mask is used.

8. What sign would an EMT see in a patient with inadequate breathing, relative to the patient's skin?
   A. Pale
   B. Cyanotic
   C. Cool and clammy
   D. All of the above

9. A patient should be suctioned immediately:
   A. When a #8.0 French suction is available.
   B. When gurgling is heard.
   C. Only during bronchospasm.
   D. When the patient has gastric distention.

10. An EMT can choose the proper size nasogastric tube by measuring from the:
  A. Tip of the nose, around the ear, to below the xyphoid process.
  B. Tip of the nose to the Adam's apple.
  C. Upper half of the sternum to the lower lips.
  D. Tip of the nose to the navel.

11. The medical term for the Adam's apple is:
  A. Cricothyroid membrane.
  B. Trachea.
  C. Vocal cords.
  D. Thyroid cartilage.

12. Intubation in infants and children must be checked following every move; in other words, from when to when?
  A. Scene to ambulance
  B. Stretcher placement to ambulance
  C. From time it is placed through entire transport to receiving facility
  D. Placement to stretcher placement

13. Placement using a straight blade ends when the tube is:
  A. Into the trachea.
  B. On top of the trachea opening.
  C. Into the area of the epiglottis.
  D. Into the area of the vocal cords.

14. When an EMT chooses the endotracheal tube for an adult, male and female patients always call for ET tubes of equal and same size.
  A. True
  B. False

15. "Murphy eye" of the ET tube decreases:
  A. Chance for vomiting.
  B. Chance of obstruction.
  C. Chance of leaks.
  D. Length of ET inserted.

16. How far should a stylet be inserted into the ET tube?
  A. To the Murphy eye
  B. To the pilot balloon
  C. 10 cm
  D. Until it enters the lungs

17. Most infant and child endotracheal tubes are uncuffed. The maximum age of a patient in which an uncuffed ET tube is used is:
  A. 4 years old.
  B. 6 years old.
  C. 8 years old.
  D. 12 years old.

18. When intubating a patient of trauma, the patient must be intubated:
  A. Slowly and meticulously.
  B. With the head and neck in a neutral position.
  C. While on their side.
  D. With the head and shoulders slightly elevated.

19. What is the concern with placing an ET tube into the abdomen?
  A. Unrecognized esophageal intubation is fatal.
  B. There is no special concern, because the abdomen is preferred.
  C. It can cause severe seizure.
  D. It must be done only when the patient is conscious.

20. Prolonged attempts while intubating may lead to hypoxia.
   A. True
   B. False

21. The narrowest part of a child's airway is the:
   A. Larynx.
   B. Epiglottis.
   C. Cricoid cartilage.
   D. Tracheal arch.

22. One of the two accepted methods for an EMT to determine proper ET tube size at an emergency scene is the size of the:
   A. EMT's little finger.
   B. EMT's thumb.
   C. Patient's little finger.
   D. EMT's ring finger.

23. What is the best method to open an airway in a patient with a suspected neck injury?
   A. Head tilt–chin lift with caution
   B. Tongue–jaw lift
   C. Jaw-thrust maneuver
   D. Jaw-thrust maneuver while the patient is in the recovery position

24. The trachea is softer and more flexible in infants and children than in adults.
   A. True
   B. False

25. Intubation must be performed quickly to remove blood, other liquids, foreign objects, and food particles.
   A. True
   B. False

Bonus Pearls

A wise man turns chance into good fortune.
— Thomas Fuller

Normal resting breathing rate for adults is _____ to _____ breaths per minute.
12; 20.

A radial pulse should be assessed in patients 12 months or older. In patients less than 12 months, a/an __________ pulse should be assessed.
Brachial.

The oxygen liter flow for a nasal cannula ranges between _____ and _____ L/min.
2; 6.

The oxygen liter flow for a medium flow (simple) mask ranges between _____ and _____ L/min.
6; 10.

The oxygen liter flow for a nonrebreathing mask should be at least _____ L/min.
15.

When palpating a patient's abdomen complaining of RUQ pain, the __________ quadrant should be palpated last.
Right upper.

If a spinal injury is suspected, the ____________ is used to bring the patient's head and neck in a neutral position.
Jaw-thrust maneuver.

Normal resting heart rate for an adult is _____ to _____ beats per minute.
60; 80.

The sudden and unexpected death of an infant in which an autopsy is unsuccessful in determining the cause of death is called:
Sudden infant death syndrome (SIDS).

Discontinuing patient care before transferring care to another health care provider with equal or greater training is called:
Abandonment.

The artery that is palpable at the top of the foot is called the:
Dorsalis pedis.

Bleeding from the nose as a result of disease, injury, or environmental factors is referred to as:
Epistaxis.

A protrusion of organs from a wound is known as a/an:
Evisceration.

A collection of blood beneath the epidermis with accompanying swelling and discoloration resulting from injury to the soft tissues is a/an:
Hematoma.

The three most important pieces of information about your patient to relay to a receiving hospital prior to arrival are __________, __________, and __________.

Age; sex; chief complaint.

A general feeling of weakness or discomfort is referred to as:

Malaise.

Any event that places excessive demands on EMS equipment and personnel is known as a/an:

Multiple-casualty incident (MCI).

Any patient suffering from shortness of breath should be transported in the __________________ position.

Full-Fowlers.

To reduce the risk of supine hypotensive syndrome, all patients in their third trimester of pregnancy should be transported in the __________________ position.

Left lateral recumbent.

A heart rate that is considered to be faster than the normal upper limit is known as:

Tachycardia.

A position commonly found with patients in respiratory distress in which the patient is sitting upright, leaning forward, while supporting the body with elbows locked is called the ______________ position.

Tripod.

A heart rate that is considered slower than the normal lower limit is known as:

Bradycardia.

Gastric distention during CPR is caused by:

Air entering the patient's stomach.

Cardiac arrest in infants and children is most commonly a result of:

Respiratory arrest.

The recovery period following the clonic phase of a grand mal seizure in which a patient, often appears weak and disoriented is called the:

Postictal state.

Signs or symptoms that might be expected based on a patient's chief complaint, but are denied by the patient, are called pertinent _________ and should be documented accordingly.

Negatives.

Low blood sugar is referred to as:

Hypoglycemia.

A bluish color of the skin and mucous membranes that indicates poor perfusion of tissue is called:

Cyanosis.

The sound or feel of broken fragments of bone grinding together is referred to as:

Crepitus.

The first priority of an EMT-B on arrival at a scene is:

Scene safety.

The state of exhaustion and irritability in the EMS provider as a result of chronic stress is referred to as:

Burnout.

The single most important way to prevent the spread of infection is through:

Hand washing.

The obligation to provide care is a concept known as:

Duty to act.

To assess breathing, use the __________, __________, and __________ method.

Look; listen; feel.

The most common cause of seizures is:

Epilepsy.

A severe form of allergic reaction is:

Anaphylaxis.

The hypotensive patient should be placed in the __________ position.

Supine.

The first method for controlling bleeding is:

Direct pressure.

During spine immobilization, manual immobilization must not be released until after the __________ is secured to the backboard.

Head.

The system used for sorting patients to determine the order in which they will receive medical care is called:

Triage.

After scene safety and body substance isolation precautions have been established, the first step in a primary survey is to:

Determine unresponsiveness.

Extremities should be assessed for ______________, ______________, and ______________ before and after application of a splint.

Circulation; sensation; motor function.

When assessing respirations, it is necessary to note __________, __________, and __________.

Rate; quality; depth.

The time it takes for compressed capillaries to fill with blood when compression is released is called ________________ time.

Capillary refill.

Central nervous system disorder or the use of narcotics may result in pupils that are:

Constricted.

Cardiac arrest or the use of drugs such as amphetamines or LSD may result in pupils that are:

Dilated.

A patient with liver dysfunction may present with a "yellowish" color evident in the __________ of the eyes.

White/sclera.

Normal resting heart rate for infants is greater than _____ beats per minute.

120.

Normal resting breathing rate for infants is greater than _____per minute.

25.

The tibia and fibula are distal to the:

Femur.

What are the three most important steps in patient assessment?

Airway; breathing; circulation.

The absence of respirations is called:

Apnea.

The appendix is in which quadrant?

Right lower quadrant.

When examining extremity trauma what three things do you check for?

Circulation; sensory; motor.

What is the normal pulse rate range for a newborn?

120–160 beats/min.

What are three contraindications for MAST?

Pulmonary edema; third-trimester pregnancy; evisceration.

According to the Rule of Nines, an adult with burns to both arms and face is what percentage?

27%.

Skin redness and sloughing of skin is what degree burn?

First-degree burn.

Weak rapid pulse, confusion, weakness, and appearance of intoxication in a diabetic are signs of what?

Hypoglycemia.

What are the four stages of labor?

Onset of contractions; crowning; delivery; placental delivery.

When crowning occurs, what is it important to check for?

Umbilical cord around the newborn's neck.

When a newborn's feet and buttocks present first, this is known as?

Breech presentation.

The portion of the sternum inferior to the body is?

Xyphoid process.

A systolic blood pressure of over 90 mm Hg, weak rapid pulse, and diaphoresis are signs of what?

Shock.

A fracture that horizontally crosses a bone is known as a/an:

Transverse fracture.

A fracture that diagonally crosses a bone is known as:

Oblique fracture.

In an extremity trauma, what are the two most important steps in management?

Splint; ice.

When a pregnant woman is in a motor vehicle accident, how should you place her on your stretcher after you have strapped her to a long backboard?

Tilt her on her left side at a 10- to 15-degree angle.

To declare a mass-casualty incident, what must you have?

Depleted the resources available to you.

How do you care for a broken tooth?

Put it back in its opening or take it out and store in milk.

In infants, where do you palpate a pulse?

Brachial artery.

What steps should you take in caring for a bleeding wound?

Pressure; elevation; pressure point; as a last resort, tourniquet.

What substances can cause anaphylactic shock?

Ingested, inhaled, or injected substances.

A tearing of skin or flapping away is called a/an:

Avulsion.

A jagged cut in the skin is called a/an:

Laceration.

The outermost layer of skin is called the:

Epidermis.

How should you handle an impaled object?

Stabilize in place and control any bleeding.

Tissue that connects bone to bone is:

Ligament.

Respiratory distress, weak rapid pulse, cyanosis, distended neck veins, and tracheal deviation are signs of:

Tension pneumothorax.

Hemothorax means what is in the pleural space?

Blood.

Congestive heart failure usually starts with failure in which portion of the heart?

Left.

A grand mal seizure consists of three phases. What are they?

Aura; tonic-clonic; postictal phase.

In children, inflammation of the larynx, trachea, and bronchi is known as:

Croup.

The APGAR scoring method checks for five things. What are they?

Heart rate; respirations; muscle tone; irritability; skin color.

You arrive on scene of a 61-year-old male patient clutching his chest, complaining of chest pain, and in respiratory distress. What method of oxygen delivery would you use?

Nonrebreathing mask.

What oxygen flow rate is required by a nonrebreathing mask?

10–15 liters per minute.

Inadequate tissue perfusion is the definition of what?

Shock.

Movement away from the body's midline is:

Abduction.

An imaginary line that divides the body into front and back sections is called:

Frontal or coronal plane.

The nearest point of origin is known as:

Proximal.

When preparing to insert an oropharyngeal airway, what other equipment should you have available to you?

Suction; bag-mask; oxygen.

Two differences between heat stroke and heat exhaustion are:

Heat stroke presents with dry hot skin and rapid pulse; heat exhaustion presents with heavy perspiration and a weak pulse.

What is the first thing you would do if you arrived on scene to an overturned tanker truck?

Scene safety and check for hazardous materials placards.

What is your primary concern on arrival on any scene?

Scene safety.

What is a normal systolic blood pressure for an adult?

100 plus the age up to 140; 10 less for females.

What is the normal pulse rate for a newborn?
120–160 beats/min.

What does the mnemonic OPQRST stand for?
Regarding pain, the Onset; Provocation; Quality; Radiation; Severity; Time.

What does the mnemonic SAMPLE stand for?
Signs and Symptoms; Allergies; Medications; Past medical history; Last oral intake; Events preceding onset.

What is a halon fire extinguisher used for?
Electronic equipment.

What is the ratio of compressions to ventilations in one-rescuer CPR?
30:20

What percentage oxygen does a BVM with supplemental oxygen deliver at 15 L/min?
100%.

How should the secondary assessment be done on a young child?
Toe to head.

True or false: The primary survey consists of airway, breathing, circulation, disability, and expose as necessary.
True.

What is hypoglycemia?
Abnormally low blood sugar.

What is anaphylaxis?
A life-threatening allergic reaction.

With medical control approval, what can the EMT use to treat a severe anaphylactic reaction?
The EpiPen auto-injector.

What drug does the EpiPen contain?
Epinephrine, also known as adrenaline.

What is a Type I ambulance?
Pick-up truck chassis with a modular box and no passage between compartments.

What is a Type II ambulance?
Van-type ambulance that may or may not have passage between compartments.

How do you measure blood pressure by palpation?
Pump up the cuff until you lose the radial pulse plus 20 mm Hg, then release the air until the radial pulse returns.

What is the maximum liter flow you can use for a nasal cannula?
6 L/min.

What is the minimum liter flow for a nonrebreathing mask?
10 L/min.

If you leave your patient in the emergency department without transferring care to a nurse or doctor, you are guilty of:
Abandonment.

What artery is compressed when you take a blood pressure on the upper arm?
The brachial artery.

When checking the pulse of an infant, where should it be checked?
The brachial artery.

What is the minimum systolic blood pressure if you can feel a radial pulse?
80 mm Hg.

Name the artery used to check circulation on an adult during the primary assessment.
The carotid artery.

How many cervical vertebrae are there?
7.

How many ribs are there in a man?
12.

What is the largest organ in the human body?
The skin.

Where is the appendix located?
The right lower quadrant.

Where is the liver located?
The upper right quadrant.

Where is the spleen located?
The left upper quadrant.

If a patient is lying on his or her left side, in which position is he or she lying?
Left lateral recumbent.

In what position is a patient lying flat on his or her back?
Supine.

Why shouldn't you place a woman who is pregnant in the supine position?
Weight of the fetus could compress the inferior vena cava.

What does CVA stand for?
Cerebrovascular accident, also known as a stroke.

When is it permitted to remove an impaled object?
When it is through the cheek and you can see both ends.

How much blood is found in an adult?
Approximately 6 liters.

What is shock?
Inadequate tissue perfusion.

Where can you check for cyanosis on a dark-skinned person?
The nail beds and gums.

What is the EMT's best defense against bloodborne pathogens?
Body substance isolation.

Where do you find phalanges?
The fingers and toes.

What is the largest bone in the body?
The femur.

What is the preferred method of controlling bleeding?
Direct pressure.

What is unique about electrical burns?
There is an entrance and exit wound.

When treating a person for electrical burns, what else should you look for?
Fractures may be present as a result of violent muscle contraction.

What does the mnemonic AVPU stand for?
Alert; responds to Verbal stimuli; responds to Painful stimuli; Unresponsive.

What is the least preferred method to control bleeding?

Application of a tourniquet.

You treat a minor who has been hit by a car, even though her parents are not at home. Consent is considered to be:

Implied.

If you treat a competent adult patient after they have told you not to touch them, you are said to have committed this legal offense:

Battery—the unauthorized touching of someone without his or her consent.

Section 1: Preparatory

1. **D** Both A and C are correct.
2. **A** Identify and treat life-threatening problems.
3. **B** First responsibility when arriving on a scene.
4. **D** Call for trained personnel to clear the scene of hazards before proceeding.
5. **B** The physician medical director of your EMS system.
6. **A** Off-line medical control.
7. **C** Medical director and public.
8. **C** You are cleaning your equipment or vehicles.
9. **B** Components of negligence.
10. **B** False
11. **D** Medical care requirements and religious choice (i.e., call a rabbi, call a priest).
12. **D** Care for the patient.
13. **A** Inferior.
14. **C** Stomach.
15. **A** Thoracic, lumbar, and sacrum.
16. **C** Tibia is referred to as the "shin bone."
17. **D** Epiglottis
18. **A** Present and equal
19. **B** Arteries
20. **C** Blood pressure cuff.
21. **D** Noisy respirations.
22. **C** Rate and quality
23. **B** Greater than 2 seconds
24. **C** Assess and record every 15 minutes.
25. **B** Altered mental status

Section 2: Airway

1. **A** Inadequate artificial ventilation
2. **B** Head tilt–chin lift maneuver
3. **C** Jaw-thrust maneuver
4. **D** Teeth
5. **B** Base of the tongue

6. C 300 mm Hg

7. C 2, 4, 3, 1

8. A 15 L/min

9. B 15 seconds at a time

10. C Body substance isolation

11. A True

12. D Infant, child, and adult

13. A Reposition the head.

14. A True

15. A True

16. B Try the other nostril.

17. D D

18. C Close the valve and bleed oxygen from the regulator.

19. C Bag is not full before placing the mask.

20. B Only when the patient will not tolerate the nonrebreathing mask.

21. A True

22. B False

23. D Adults and children

24. C It may cause instant vomiting and aspiration.

25. B False

Section 3: Patient Assessment

1. C Scene safety

2. D The patient and bystanders

3. B Begin triaging patients.

4. C Assessment to ensure the well-being of the EMT.

5. B Airway and mental status

6. B Look, listen, and feel for adequate breathing.

7. A Over the radial pulse.

8. C Palpate the carotid pulse.

9. C Nail beds, lips, and eyelids.

10. A 2 seconds or less

11. A True

12. B SAMPLE history

13. C From bystanders, friends, or family

14. B Drainage and bleeding

15. B Tell medical control that you question the order.

16. B False—courtesy is assumed.

17. C "Affirmative" or "negative"

18. C Both A and B are correct.

19. A Be honest with your patient.

20. D Elderly patient

21. C Act and speak in a calm, competent manner.

22. B In the minimum data set.

23. D Document what did or did not happen and what steps, if any, were taken to correct the situation.

24. A True

25. B Prepare a special-situation report.

Section 4: Medical Emergencies

1. C Situations in which a drug should not be used because it may cause harm to the patient

2. A Oxygen.

3. B 15–30 breaths per minute.

4. A Stridor

5. A Being highly agitated and aggressive.

6. B Provide supplemental low-flow oxygen.

7. D Patient must be seizing and over 14 years old.

8. C 25–50 grams

9. C Postexercise dyslexia

10. A The formation of blood clots

11. B Lack of oxygen or low blood sugar (diabetes).

12. B Suicide.

13. A Its own conduction system

14. C Both A and B are correct.

15. D Capillaries

16. A True

17. B False

18. C 2, 4, 1, 3

19. B Contact medical control (medical direction).

20. C Between the clothing and the patient's abdomen.

21. A Reduces the body's ability to lose heat through evaporation

22. D On the left side.

23. B Place the pad to allow fluids to drain.

24. B 10 minutes

25. D 100; 60

Section 5: Trauma

1. D 2, 4, 3, 1

2. C Apply and maintain mechanical traction.

3. A Delivers oxygen and other nutrients to the cells of all organ systems

4. D Arterial

5. B 4 inches wide.

6. A True

7. D Internal bleeding.

8. B Cell and organ death

9. C Skin.

10. C Avulsion

11. D Large amounts of water for at least 20 minutes.

12. B Ensure that the scene is safe to enter.

13. C Vital signs

14. C 10–20 percent

15. B False

16. A Increased pulse rate (early sign) and decreased blood pressure (late sign)

17. C Concern for reattachment

18. A Pulling apart of the spine.

19. B Mastoid process.

20. B Maintain manual immobilization until the head is secured to the board.

21. C 1. Start manual in-line immobilization; 17. Reassess pulses, motor, and sensation and record.

22. B False

23. A True

24. A The patient's airway and breathing.

25. C Control bleeding.

Section 6: Infants and Children

1. D Increasing the effort to breathe.

2. A Rapid respiratory muscle fatigue.

3. C Visual foreign body removal and abdominal thrusts.

4. A Squeeze the bag slowly and evenly so the chest rises adequately.

5. B Skin temperature

6. D Respirations and skin condition

7. D Both A and B are correct.

8. C Respiratory status.

9. B False

10. C Always.

11. A Seizures.

12. B 10

13. B Protecting the child from injury.

14. D Partial airway obstruction

15. C En route to the hospital

16. B Begin artificial ventilation.

17. C Sudden infant death syndrome

18. C Blunt trauma

19. D Possible femur fracture

20. C Modified jaw thrust.

21. A True

22. B An altered mental status that is not related to hypoglycemia.

23. C It can be done safely.

24. A Proper function of the brain.

25. D Itchy, watery eyes

Section 7: Operations

1. C 24-hour availability; trained personnel.

2. A That they are properly buckled up

3. B A safe entrance to traffic and pay attention to weather and road conditions.

4. D Headlights.

5. B The hazards to the crew to increase considerably.

6. A True

7. B False

8. B Multiple vehicle response.

9. C Crashes at intersections.

10. B Shut off the headlights unless they are necessary to illuminate the scene.

11. A True

12. A Notify dispatch.

13. **D** Nonlatex gloves, gowns, and eyewear.

14. **B** Call dispatch with an accurate request for additional help.

15. **B** Notify dispatch.

16. **C** Administer necessary care to the patient before extrication and ensure that the patient is extricated in a way that minimizes further injury.

17. **D** Safety

18. **A** Verbal; written

19. **B** Notify dispatch.

20. **B** Delayed movement would endanger the life of the patient and rescuer.

21. **D** All of the above

22. **D** All of the above

23. **C** Environmental conditions and hazards

24. **B** Sector officer.

25. **C** The most knowledgeable EMT arriving on the scene

Section 8: Advanced Airway

1. **B** Epiglottis.

2. **A** Thoracic cavity

3. **B** Right intercostal space.

4. **D** Cricoid cartilage

5. **A** Adequate and equal.

6. **D** Just before death.

7. **B** There is suspicion of a neck injury.

8. **D** All of the above

9. **B** When gurgling is heard.

10. **A** Tip of the nose, around the ear, to below the xyphoid process.

11. **D** Thyroid cartilage.

12. **C** From time it is placed through the entire transport to receiving facility

13. **A** Into the trachea.

14. **B** False

15. **B** Chance of obstruction.

16. **A** To the Murphy eye

17. **C** 8 years old.

18. **B** With the head and neck in a neutral position.

19. **A** Unrecognized esophageal intubation is fatal.

20. **A** True

21. C Cricoid cartilage.

22. A EMT's little finger.

23. C Jaw-thrust maneuver

24. A True

25. B False